BRENTWOOD
D0397624

ALSO BY JEFFREY BENNETT:

TEXTBOOKS

Life in the Universe

The Cosmic Perspective

The Essential Cosmic Perspective

Using and Understanding Mathematics

Statistical Reasoning for Everyday Life

BOOKS FOR CHILDREN

Max Goes to the Moon

Max Goes to Mars

Max's Ice Age Adventure (with Logan Weinman)

Max Goes to Jupiter (coming soon)

BOOKS FOR ADULTS

On the Cosmic Horizon

BEYOND UFOS

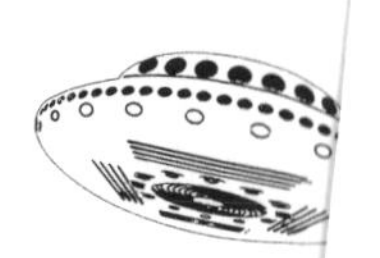

BEYOND UFOS

The Search for Extraterrestrial Life and Its Astonishing Implications for Our Future

JEFFREY BENNETT

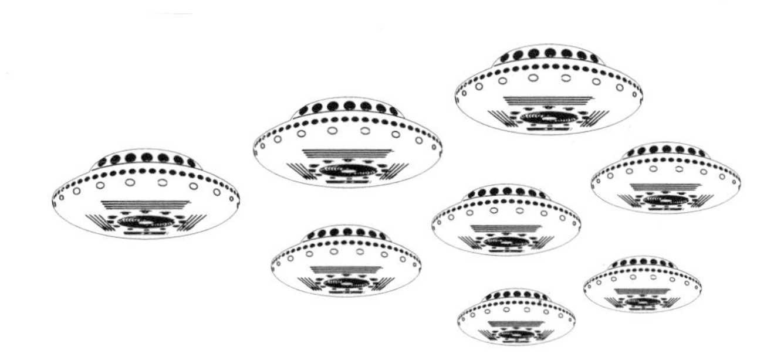

CONTRA COSTA COUNTY LIBRARY

PRINCETON UNIVERSITY PRESS
PRINCETON AND OXFORD

3 1901 04389 4460

WITHDRAWN

Copyright © 2008 by Jeffrey Bennett
Requests for permission to reproduce material from this work should be sent to Permissions, Princeton University Press

Published by Princeton University Press, 41 William Street, Princeton, New Jersey 08540

In the United Kingdom: Princeton University Press, 6 Oxford Street, Woodstock, Oxfordshire OX20 1TW

All Rights Reserved

ISBN-13: 978-0-691-13549-5

Library of Congress Cataloging-in-Publication Data

Bennett, Jeffrey O.
Beyond UFOs : the search for extraterrestrial life and its astonishing implications for our future / Jeffrey Bennett.
p. cm.
Includes index.
ISBN 978-0-691-13549-6 (alk. paper)
1. Exobiology. 2. Life on other planets.
3. Life—Origin. I. Title.
QH327.B446 2008
576.8'39—dc22 2007037872

British Library Cataloging-in-Publication Data is available

This book has been composed in Aldus
Printed on acid-free paper. ∞
press.princeton.edu
Printed in the United States of America
1 3 5 7 9 10 8 6 4 2

To my wife, Lisa,
and my beautiful children,
Grant and Brooke

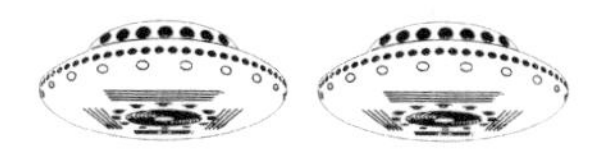

May your lives be filled with laughter
as you reach for the stars.

All this world is heavy with the promise of greater things, and a day will come, one day in the unending succession of days, when beings, beings who are now latent in our thoughts and hidden in our loins, shall stand upon this earth as one stands upon a foot-stool, and shall laugh and reach their hands amidst the stars.

—H. G. Wells, 1902

CONTENTS

PREFACE

Alien Dreams

It has been a long time, but I distinctly recall the sounds and sights of my alien friends. I was only about 8 or 9 when they first visited me. I frequently heard their voices calling to me, as small sounds in the night, like whispers into my ear. I'd listen hard, trying to make out a few words. Sometimes I even saw them. They signaled with little flashes as the dawn light entered my room, or in daytime with dancing lights in rays of sunshine. They must have been quite tiny, these friends from other worlds. At night, I'd look for signs of their spaceships in the sky. I knew that someday they would invite me to take a trip with them, carrying me far from adolescent troubles to wondrous places of beauty, adventure, and joy. Out there, pain and suffering were unknown. They would show me the galaxy, they would choose me as their ambassador to Earth, and I would bring the news of our great new future to all of humanity. It was only a matter of time, and of my ability to decipher their whispers in the night.

I'm now in my fifth decade of life, and I have not seen or heard from my friends since I was a teenager. I still believe they are out there, somewhere, but I am no longer convinced that they really visited me in my bedroom. Growing up can change your perspective in remarkable ways. The tooth fairy and Santa Claus don't seem quite the same now as they did when I was a child. My little stuffed bear, whom I'd affectionately named Wilshire Boulevard, never did grow back the hair I clipped off him one day, even though I was sure he would. My alien visitors seemed so very real, but now it seems more likely that I saw reflections from dust suspended in the air, and that the whispers in the night were just the sounds that I still hear but now interpret quite differently.

Still, I have not given up hope of someday meeting my friends again. I am more convinced than ever that wondrous worlds really do await us, if only we can put aside our human hatreds and wars and survive long enough to meet those who have inhabited this universe for millions or billions of years before us. Yes, growing up can change your perspective in remarkable ways. I no longer believe in the voices in the night, in the dancing lights in rays of sunshine, or in the world of magic. Instead, I now *know* that my imagination was far too limited, for the real world holds wonders far more mystifying and incredible than anything my young mind could ever have conjured up.

My personal journey from wide-eyed child to wide-eyed scientist was inspired not only by my alien friends but also by two real events of the late 1960s. The most important by far occurred on July 20, 1969, when Neil Armstrong and Buzz Aldrin became the first human beings to set foot on another world. The fact that the Apollo Moon landings occurred during a time of great turbulence, with both the Vietnam War and the Cold War in full swing, only amplified their message of hope. For myself and other young children of the era, Apollo told us that the troubles of the present would eventually give way to a future that held new worlds. The other event was the release of the movie *2001: A Space Odyssey*. This movie, which opened a year before Apollo reached the Moon, was my first exposure to the idea that the universe might have beings so advanced that they would seem incomprehensible to us. I became a science fiction aficionado, which allowed me to continue journeying into space even as the Apollo program wound down. It's been over 35 years since the last real people walked on the Moon in December 1972, but at night my dreams routinely allow me to travel far beyond.

Between the movie, the Apollo reality, and my own alien friends, it's no accident that I decided to pursue a career as a space scientist. Still, it was hardly a straight-line path. I started college as an engineering major, thinking I might build spacecraft, but soon decided I was more interested in the places that spacecraft might visit than in the rockets themselves. By the time I graduated (from the University of California at San Diego) I had become more pragmatic, majoring in biophysics because it seemed to offer more direct benefits to society than studying distant stars and galaxies. Luckily for me, Carl Sagan's *Cosmos* series aired just in time to show me the error of my ways.

Sagan's series is still remarkably fresh despite being more than two decades old, and it beautifully illustrates the reason I altered my own career path: It shows us that knowledge of the cosmos is not just idle knowledge but rather is at the core of everything that makes us human. I took the message to heart and sent letters turning down the schools that had offered me graduate study in biophysics. I had already been very interested in teaching, having worked through college as an elementary school teaching aide (grades 2–3), and decided that my calling would be to try to follow Sagan's footsteps as a popularizer of astronomy. I ended up at the University of Colorado in Boulder, because their program was flexible enough to allow me to pursue my interest in teaching while still obtaining a doctoral degree in astrophysics. Thanks to my ever-patient thesis advisor, Tom Ayres, I was even allowed to spend time nearly equivalent to what I spent on my thesis in pursuing an educational project that involved building a scale model of the solar system (discussed in chapter 3).

I don't believe in fate, but with hindsight it's rather striking how my detours set me up perfectly to be writing about aliens today. Largely because I was one of the few people who actually volunteered to teach freshmen courses, I ended up teaching several courses each year throughout my graduate student career. My love of teaching and my work on the scale model solar system caused my name to come up in a faculty committee charged with helping to revise the university's core requirements, and the timing worked out just right for me to become the director and lead teacher for a new mathematics program for liberal arts students, a program that we based on the idea that the ability to think critically about numbers and mathematical ideas—what we call "quantitative reasoning"—is for most students much more important than the equation-solving skills taught in traditional mathematics courses such as college algebra. The lack of existing material for the new course caused me to start writing my own, which soon led me to a publishing contract for a college mathematics textbook. When the publisher found out that I was actually an astronomer rather than a mathematician, I was offered my next contract, this time for a college astronomy textbook. Both books proved fairly successful, setting me up for the fateful day when my astronomy editor, Adam Black, asked me if I'd consider writing a textbook about the search for life in the universe.

I said no, of course, because I didn't feel I knew enough about the subject. Unlike the subjects of my prior textbooks, this was a course I'd never taken and never taught. However, Adam promised to team me up with some bona fide experts in the field, and with a little study I realized that my undergraduate background in biology would actually come in useful. The idea that my past detours would suddenly be of value was too much to let go of. I agreed to the project and Adam arm-twisted two of the world's foremost astrobiologists into working with me (Bruce Jakosky from Boulder, who left the project after the first edition, and Seth Shostak of the SETI Institute, who continues to work with me). I often felt like I was an undergraduate student again as they patiently tried to explain the science to me. I'm not always quick to grasp new things, and I probably asked more "stupid questions" of them than all my students combined have ever asked of me. Nevertheless, I eventually learned enough so that we could successfully complete the textbook, called *Life in the Universe,* which is now in its second edition and has become the leading college textbook for introductory-level courses in the young field that NASA calls "astrobiology."

And you know what I was thinking about the whole time I worked on that project? Every time I saw rays of sunlight shining in through my window, I thought about my long lost alien friends. I realized that with my new understanding, I could finally start to think about them again, but this time with an eye toward real science instead of just dreams. I also realized that the questions raised by the search for life in the universe go far deeper than I had naively expected. They touch on issues of the origin of life, the origin of intelligence, the nature of the human mind, and the survivability of our civilization. Most relevant to the task at hand, I decided that these issues cut so deeply that both individuals and cultures might change for the better if all were aware of them. I decided that I should take what I'd learned in writing a college textbook and present the crucial lessons in a format that could be read by anyone, in hopes of sharing my newfound understanding with as many people as possible. If you are reading these words, then I know I've succeeded at least in getting you through the first few pages.

If you are willing to continue on, this book will take you on a short journey through the world of science and what it has to say about the possibility of life beyond Earth. I will explain why life elsewhere seems ever more likely, why many scientists suspect that civilizations are also common, and the surprising things that we can say about alien visitors to Earth, even while I am personally skeptical of claims that they are here. I'll describe how the search for life on other worlds is helping us understand life on Earth, illuminating the remarkable circumstances of our planet and showing us how we may be unwittingly threatening our own existence. I'll discuss the search for life elsewhere, both within our own solar system and beyond. Most important, I'll tell you why I believe that the quest to find life beyond Earth may help us overcome the ailment that I call *center of the universe syndrome*—the syndrome that makes too many people behave as though the universe really does revolve around them. So with that introduction, let's leave our alien dreams behind, and look into what science really has to say about the search for extraterrestrial life and its astonishing implications for our future.

ACKNOWLEDGMENTS

This book carries only my name as author, but in fact it is the result of collaborative work with many other people. First and foremost, I'd like to thank my textbook co-authors—especially Megan Donahue, Nick Schneider, Mark Voit, and Seth Shostak—who have helped me at some point with nearly all of the topics discussed in this book. Indeed, they'll even recognize some of the exact words in the book as ones we've worked on together and that I borrowed from our textbooks. Many other friends and colleagues—too many to name—have also contributed through the great discussions we've had over the years.

Extra special thanks also to my good friend Joan Marsh, who has worked on nearly every book I've ever written and carefully read this manuscript multiple times, suggesting many changes and helping me get it to the point where it could impress a publisher enough to make it into print. Thanks also to my agent, Skip Barker, and editor Ingrid Gnerlich, who turned the manuscript into a published book.

Finally and most importantly, I thank my wife, Lisa, without whom I would never have had the opportunity to write any of my books. A little over a decade ago, I was making a living by teaching undergraduate courses and doing grant-supported research. Lisa made my transition to writing possible by supporting us almost single-handedly during the many years it took to write the first editions of my textbooks, which ultimately led to this and other books as well. She has also been my most trusted advisor and best friend, and she is the kindest, most intelligent, and most thoughtful person I have ever had the privilege to know.

BEYOND UFOS

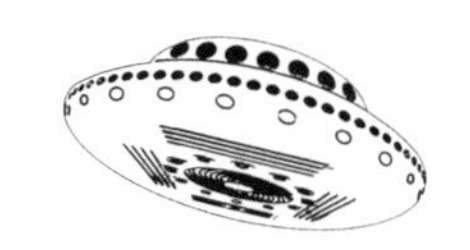

1
WORLDS BEYOND IMAGINATION

> Do there exist many worlds, or is there but a single world? This is one of the most noble and exalted questions in the study of Nature.
>
> —*Saint Albertus Magnus (c. 1206–1280)*

This is a book about possibilities. It is about the possibility that, within a decade or two, robotic or human explorers will drill into the Martian surface and discover microscopic life in subterranean pockets of liquid water. It is about the possibility of landing spaceborne submarines on Jupiter's moon Europa, where they might melt their way through miles of ice and observe life swimming in a volcanically heated ocean. It is about the possibility of strange, cold-adapted life forms on Saturn's moon Titan, a world on which we have already landed a robotic emissary, despite its being located nearly a billion miles away. It is about the possibility of SETI researchers detecting an unmistakable signal coming to us from a civilization that has grown up around a faraway star. It is about the possibility that we may already be surrounded by a galactic civilization, populated by beings who surpassed our own current level of development millions or even billions of years ago. Most of all, it is about the possibilities that await *us,* if and when we learn that we are not alone in the universe.

It doesn't take long to begin to appreciate these and other possibilities, but you have to be in the right frame of mind. If you're reading at night and you happen to live in a place with clear, dark skies, take a moment to put the book down and go out and look at the stars. If you live in a city or it is cloudy or daytime, close your eyes and picture yourself at a favorite vacation spot on a perfect night. Personally, I like the mountain lakes not far from my home in Colorado, where the stars sometimes shine so brightly that I can make out the constellations by their reflections in the still water.

As you look out into the seemingly infinite heavens, you should feel a change in your mental state as your thoughts shift from the daily trials of life to questions of who we are, how we got here, why we exist, and whether we have companionship among the planets and stars.

The mere sight of the myriad stars may seem enough to answer the last question. After all, when you consider the fact that each star is a sun, possibly orbited by planets of its own, it may seem inevitable that others are out there, looking at us as a dot of light in their own skies. But possibilities are not certainties, and despite everything we know about the universe today, we still have no proof that even the tiniest microbes live beyond the confines of our small world. We may have good reason to be entranced by the possibilities for life beyond Earth, but it is also possible that such life exists nowhere except in our own minds.

That is where science comes in. Science is a way of distinguishing possibilities from realities. We can imagine all the possibilities that we want, but science asks us to put them to the test. If we find confirming evidence for our possibilities, then we have at least some reason to think they reflect reality. If our possibilities conflict with reality, then we know they were figments of our imagination. Of course, oftentimes we have no clear evidence either way, as is the current case for the possibility of extraterrestrial life. In such cases, the job of science is to help us keep looking and learning, until we someday acquire the evidence we seek.

Today, many hundreds of scientists around the world are engaged in the scientific search for life in the universe, a topic of study that is often called *astrobiology* or *exobiology*. In the United States, NASA has established an Astrobiology Institute, which functions as a collaborative effort between scientists at NASA research centers and at more than a hundred universities and independent research laboratories. The European Union has a similar collaborative effort with its European Exo/Astrobiology Network. Australia, Great Britain, Spain, France, and Russia also have formal astrobiology centers, and almost every other nation on Earth has at least a few scientists whose research bears on the question of life in the universe.

Given that we don't yet know of any life beyond Earth, you might wonder how so many scientists can be gainfully employed in its study. The answer, like this book, is about possibilities. Only a few scientists—those involved in the search for extraterrestrial intelligence, or SETI for short—are currently engaged in a direct effort to detect alien life. For all the others, current efforts focus on learning about the possibility of life existing elsewhere. For example, planetary scientists explore other worlds in our solar system either telescopically or by sending out robotic spacecraft. While their efforts could in principle turn up direct evidence of life, for the time being they are more focused on helping us understand the conditions found on different worlds, thereby allowing us to evaluate whether those conditions might be conducive to life. Many scientists working in astrobiology study the basic chemistry and nature of life, which should help us recognize alien life if we happen to come across it. Others seek to understand the origin of life on Earth; after all, an understanding of how life arose on our own planet ought to make it easier for us to determine the likelihood that life might arise somewhere else. Still others study Earth itself, which teaches us about how the geological nature of Earth helps make it home to abundant life. Even astronomers get in the game, seeking stars that could make good suns, looking for planets around those stars, and developing technologies that may someday help us detect life even on worlds that we can study only through telescopes.

Of course, all this effort is predicated on the idea that the possibility of extraterrestrial life is worthy of scientific study. Here, we must distinguish between an idea that is *philosophically reasonable* and one that is *scientifically testable*. The fact of our own existence makes it philosophically reasonable to wonder if life exists beyond Earth, but until quite recently there was no way in which we could actually test out the idea. In most of the rest of this chapter, I will try to explain why, in just the past couple of decades, the search for life in the universe has suddenly become a topic of intense scientific interest. First, however, it's worth developing a bit of historical perspective on the philosophical question that drives us to wonder if we are alone.

THE ANCIENT QUESTION OF WORLDS BEYOND EARTH

Even aliens need a place to call home. No matter whether we consider the tiny intelligent beings who I once imagined visiting my bedroom or the most primitive single-celled slime, all life must have gotten its start somewhere. Thus, the question of life beyond Earth makes sense only if we have reason to think that there are other worlds upon which life could live.

Those of us who would like to meet aliens generally take it for granted that the universe is indeed full of hospitable planets on which life and civilizations might have arisen. We cannot yet be certain that this is the case, because our technology is not quite yet up to the task of discovering such planets around other stars. Nevertheless, as I'll discuss in more detail shortly, the idea seems reasonable today because we know that other stars have at least some planets, and our understanding of planetary formation makes it plausible to imagine that planets with life could turn out to be common. But if we go back just a few centuries, the context for considering life beyond Earth was quite different.

Consider the quotation from Saint Albertus Magnus that opens this chapter, which begins: "Do there exist many worlds, or is there but a single world?" If you read Magnus's quotation with a modern eye, you might think he's using the term *world* in the sense of an Earth-like world with life. But he was actually using it in a much more basic way. Before the time of Copernicus, Kepler, and Galileo, all of whom lived less than 500 years ago, scholars generally assumed that Earth held a central place in the universe. Our solid home—which, by the way, had been known to be spherical since the time of the ancient Greeks—was assumed to be surrounded by a great sphere of stars, and between Earth and the stars lay additional spheres that carried the Sun, the Moon, and the five planets known at the time. Thirteenth-century philosophers and theologians had no more reason to think of any of these objects as "worlds" than to think of them as gods—an idea that had long since been rejected as ancient mythology.

In fact, pre-Copernican scholars did not even consider Earth to be a planet. The word *planet* comes from the Greek for "wanderer," and it originally referred only to objects that appear to wander among the stars in our sky. The idea will be clear if you think about the universe as it appears to the naked eye. We live on our seemingly central and unmoving home, while the stars appear to circle around us with each passing day, always staying in the fixed patterns of the constellations. The Sun also makes a daily circle around us, but not quite at the same rate as the stars. That is why the Sun gradually makes its way through all the constellations of the zodiac over the course of a year. The Moon follows this same basic pattern of motion, but moves more quickly through the constellations than the Sun, completing a full circuit and cycle of phases in about a month (think "moonth"). Before the era of airplane lights and aside from an occasional comet, the only other objects that ever seemed to move against the background of the stars were the five bright points of light known as Mercury, Venus, Mars, Jupiter, and Saturn. Thus, from the perspective of people living more than 500 years ago, there were seven objects that appeared to wander among the stars and hence qualified as "planets": the Sun, the Moon, and the five innermost planets besides Earth. The planetary status of these seven objects is enshrined in the names of the seven days of the week.[1] In English, only *Sun*day, *Moon*day, and *Saturn*day are obvious, but

[1] Want an example of how deeply astronomy is intertwined with our everyday lives? Just think about the fact that the planet Uranus *is* faintly visible to the naked eye, and that if ancient people had noticed it wandering relative to the stars we probably would have had 8 days a week instead of 7.

if you know a romance language like Spanish you'll be able to figure out the rest: Tuesday is Mars day (martes), Wednesday is Mercury day (miércoles), Thursday is Jupiter day (jueves), and Friday is Venus day (viernes).

Because the Earth-centered belief system implied that our world should be fundamentally different from any of the lights in the sky, you might wonder how Saint Albertus Magnus could even have conceived of other worlds. The answer is that, following a line of thought dating back to ancient Greece, he was considering the possibility of other worlds that were more like what we might think of as separate universes—each world the center of its own cosmos, circled by its own sun, planets, and stars. The question he asked also dated back to the ancient Greeks, inspired by his reading of Aristotle, which at the time had recently been translated into Latin.

It can be tempting to think that people who lived more than 2,000 years ago were more primitive or simpleminded than we are, but in fact many ancient civilizations were remarkably sophisticated. The ancient Greeks, geographically positioned at a crossroads that gave them access to ideas and inventions from cultures throughout Eurasia and northern Africa, developed philosophies that still resonate today. On the question of other worlds and extraterrestrial life, the Greeks split into two distinct camps.

On one side were the *atomists,* Greek philosophers who held that everything is made of tiny, indivisible atoms of four basic elements: fire, water, earth, and air. The atomist doctrine was developed largely by Democritus (c. 470–380 B.C.), who argued that the world—both Earth and the heavens—had been created by the random motions of infinite atoms. For example, he imagined atoms of earth to be rough and jagged, like tiny pieces of a three-dimensional jigsaw puzzle, so that they could stick together when they collided and thereby explain how our world had formed in the first place. Because the atomists believed the total number of atoms to be infinite, they assumed that the same processes that created our world should also have created others. This inevitably led them to conclude that other worlds and other life must exist, an idea summarized in the following quotation from the atomist philosopher Epicurus in about 300 B.C.: "There are infinite worlds both like and unlike this world of ours. . . . we must believe that in all worlds there are living creatures and plants and other things we see in this world."[2]

[2] From Epicurus's "Letter to Herodotus"; I found both this quotation and the next one from Aristotle in David Darling's wonderful reference book, *The Extraterrestrial Encyclopedia* (New York: Three Rivers Press, 2000).

Although it's difficult to ascribe modern sentiments to ancient beliefs, the atomists seem to have been essentially atheistic. They did not see the need for any hand of God in creation, instead just seeing random events in infinite time and space. However, in the pre-Christian era it was not the question of God that bothered their detractors so much as the question of infinity.

Aristotle (384–322 B.C.) represented the opposing camp. Like the atomists, Aristotle assumed the world to be made of the four elements, fire, water, earth, and air. But he did not necessarily accept that these elements could be broken down into indivisible atoms, and he certainly didn't agree that they floated randomly in an infinite space. Instead, Aristotle held that all elements had their own natural motion and place. For example, he believed that the element earth moved naturally toward the center of the universe, an idea that offered an explanation for the Greek assumption that Earth resides in a central place. Water, being lighter, settled on top of earth, thus explaining oceans, while air settled above that to explain the atmosphere. The element fire, he claimed, naturally rose away from the center, which is why flames jut upward into the sky. These incorrect ideas about physics, which were not disproved until the time of Galileo and Newton almost 2,000 years later, caused Aristotle to reject the atomist idea of many worlds. If there were more than one world, there would be more than one natural place for the elements to go, which would be a logical contradiction. Aristotle concluded: "The world must be unique. . . . There cannot be several worlds."

Aristotle also came to a very different conclusion than the atomists about the nature of the sky. Because he had natural places for all four elements to go, he concluded that the heavens must be made of something else, which he called the *ether* (literally, "upper air"). That's how the word *ethereal* came to mean "heavenly." You may also recognize that the ether was in a sense a fifth element after fire, water, earth, and air, thus explaining how the word *quintessence*—which literally means "fifth element"—came to be associated with heavenly perfection.

Interestingly, Aristotle's philosophies were not particularly influential until many centuries after his death, when his books were finally translated into Latin and came to the attention of people like Saint Albertus Magnus and one of his students, Saint Thomas Aquinas (1225–1274). Aquinas found Aristotle's philosophy particularly appealing and integrated it into Christian theology. The contradiction between the Aristotelian notion of a single world surrounded by heavens and the atomist notion of many worlds in an infinite universe became a subject of great concern to Christian theologians. Many even argued that extraterrestrial life could not be possible because it would contradict the Aristotelian notions of Earth and heaven. While a few biblical fundamentalists still take this position, it's fairly clear that the Bible itself does not weigh in on the question of life beyond Earth. As a result, today you can find fundamentalist Christians who also believe in UFOs.

From my own standpoint, the most fascinating part of this historical debate is that it continued for some two thousand years and led many people to question the very foundations of theology, even though it not only lacked any facts to back it up but was based on something that we now know to be patently untrue: Earth is *not* the center of the universe, after all. You might think this would have been a lesson learned for later generations, but sadly, we humans never learn quite so easily.

IF ARISTOTLE WAS WRONG . . .

In 1543, Nicholas Copernicus published *De Revolutionibus Orbium Coelestium* ("Concerning the Revolutions of the Heavenly Spheres"), a book in which he made the radical suggestion that Earth was not in fact the center of the universe, but instead was one of the planets going around the Sun. It was not an entirely new idea; some 1,800 years earlier, a Greek philosopher named Aristarchus (c. 310–230 B.C.) had proposed the same thing, and Copernicus was aware of Aristarchus's work when he wrote his book. However, while Aristarchus had little success in convincing any of his contemporaries of the idea's validity, Copernicus started a revolution. It took a few decades and the help of people like Tycho Brahe, Kepler, and Galileo, but by the mid-1600s the idea of an Earth-centered universe was essentially dead.

The death of the Earth-centered idea had many profound, philosophical implications. Among other things, it forced a redefinition of the word "planet": Instead of being something that moved relative to the stars in our sky, it came to mean an object that orbits the Sun. Placing Earth among the planets also provided the first actual evidence with which scientists could evaluate the ancient debate between Aristotle and the atomists, and the verdict couldn't have been more clear: Aristotle was wrong, because his entire argument for Earth's uniqueness had been based on the suddenly discredited idea that it was located at the center of the universe.

Of course, the fact that Aristotle was wrong did *not* automatically mean that the atomists had been right, but many of the Copernican era scientists assumed that they had been. Galileo suggested that lunar features he saw through his telescope might be land and water much like that on Earth. Kepler agreed and went further, suggesting that the Moon had an atmosphere and was inhabited by intelligent beings. Kepler even wrote a science fiction story, "Somnium" ("The Dream"), in which he imagined a trip to the Moon and described the lunar inhabitants.

Later scientists took the atomist belief even further. William Herschel (1738–1822), most famous as co-discoverer (with his sister Caroline) of the planet Uranus, assumed that all the planets were inhabited. In the late nineteenth century, Percival Lowell famously imagined seeing canals on Mars, attributing them to an advanced Martian civilization, an idea that led H. G. Wells to write *The War of the Worlds.*

If all this debate about extraterrestrial life shows anything, it's probably this: *It's possible to argue almost endlessly, as long as there are no actual facts to get in the way.* With hindsight, it's easy for us to see that everything from the musings of the ancient Greek atomists to the Martian canals of Percival Lowell were based more on hopes and beliefs than on any type of real evidence.

Nevertheless, the Copernican revolution really did mark a turning point in the debate about extraterrestrial life. For the first time, it was possible to test one of the ancient ideas—Aristotle's—and its failure led it to be discarded. And while the Copernican revolution did not tell us whether the atomists had been right about life, it did make clear that the Moon and the planets really are other *worlds,* not mere lights in the sky. This fact alone made it plausible to imagine life elsewhere in our solar system, even if we still knew little about the nature of those worlds.

THE NATURE OF WORLDS

The post-Copernican optimism regarding life on other worlds of our solar system never fully subsided, as even today we regard a few places—such as Mars, Europa, and Titan—as potential homes for life. Nevertheless, scientific enthusiasm for life in our solar system dampened significantly during much of the twentieth century. Improvements in telescopic technology gave us better images of the Moon and planets, and scientists learned to use techniques of spectroscopy—the dispersal of light into a rainbow-like spectrum—to learn about the composition and other properties of distant worlds.

Images and spectra quickly ruled out the idea of oceans and atmosphere on the Moon, and it likewise became clear that Lowell's Martian canals simply did not exist. Spectroscopy helped scientists discover that Venus is a searing hothouse, making life of any kind seem highly unlikely. By the mid-1960s, the advent of the space age had brought us our first close-up images of Mars, revealing a landscape littered with craters. Not only was there no sign of civilization, but the absence of liquid water made prospects look bleak even for much simpler forms of Martian life. Other worlds offered little more encouragement, as we soon realized that, in our solar system at least, surface liquid water is unique to Earth.

The Copernican revolution also opened the possibility of life among the stars. Once we learned that stars are distant suns, it seemed plausible to imagine that other stars could have their own planets, perhaps with life. However, even this idea suffered during the first half of the twentieth century, a time during which many scientists thought our solar system might have been created by a rare near-collision between stars. Calculations showed that if our planetary system was born in such a stellar collision, the odds were long against there being even a single other planetary system among the stars visible in the night sky. Prospects of life within our solar system looked dim, and prospects of worlds beyond seemed even dimmer. No wonder that scientists in the mid-twentieth century paid fairly little attention to the search for life beyond Earth.

So what changed to make extraterrestrial life such a hot topic of scientific research today? A lot. As we learned more about our own solar system, we began to realize that other planetary systems probably are *not* uncommon, making it seem much more reasonable that other stars could have Earth-like planets. Moreover, while we now have enough spacecraft images to say confidently that no other world in our solar system has ever been home to a civilization, we've also learned that at least a few worlds have conditions that might allow for life of some kind. At the same time, astronomers began to get a real handle on the size and age of the universe, demonstrating not only that there must be an enormous number of worlds on which life might have arisen, but also that there has been plenty of time for life to arise and evolve. Meanwhile, as biologists learned more about the nature of life on Earth, we began to realize that humans and other animals are not really "typical" of most life. Instead, most life is microscopic, and lives under conditions that would seem quite alien to us—so alien that it suddenly became plausible to imagine life surviving under the harsh conditions of places like Mars. Let's discuss these ideas in a little more depth, so that you will understand why, here at the dawn of the third millennium, it seems eminently reasonable to imagine that we'll soon discover life beyond Earth.

THE PLANETARY CONTEXT

Science often progresses in fits and starts, and the question of the origin of our solar system is a good case in point. Today, scientists think that our solar system formed from the gravitational contraction of a giant cloud of gas and dust floating in interstellar space. This basic idea was first proposed in 1755 by the German philosopher Immanuel Kant (1724–1804). About 40 years later, French mathematician Pierre-Simon Laplace (1749–1827) put forth the same idea independently.

According to the idea of Kant and Laplace, the Sun and the planets formed naturally as a result of processes that should occur in any collapsing cloud of interstellar gas. Their idea therefore leads almost automatically to the conclusion that other stars should have formed similarly to our own Sun and should be similarly surrounded by planets. However, while Kant and Laplace had an idea that we now believe to be correct, they were unable to back their idea with much in the way of evidence. Moreover, Laplace proposed a specific mechanism by which he claimed the planets were made; by the early twentieth century, other scientists had concluded that the mechanism could not really work as Laplace had thought.

With no real evidence to back the Kant-Laplace hypothesis and at least some reason to think that it could not work, early-twentieth-century scientists sought alternate explanations for the birth of our solar system. Many began to favor an even older idea: In 1745, ten years before the publication of Kant's hypothesis, French scientist Georges Buffon (1707–1788) suggested that the planets had been born when a massive object collided with the Sun and splashed out debris that coalesced into the planets. In the twentieth-century version of Buffon's idea, a direct collision was no longer necessary; instead, scientists imagined that the planets formed from blobs of gas that were gravitationally pulled out of the Sun during a near-collision with another star. As I noted earlier, the near-collision idea would have had dire consequences for the possibility of finding other Earth-like planets and life, because it would have meant that planets could form only in exceedingly rare events rather than as a natural part of the star formation process.

The ascendance of the near-collision hypothesis caused scientists to study it in much more depth, and to try to work out the precise physics by which the planets would have formed. As the calculations improved, the near-collision idea began to run into problems similar to those that had earlier plagued the Kant-Laplace hypothesis. In particular, try as they might, scientists could not come up with any way by which a near-collision could explain either the precise orbits of the planets in our solar system or the fact that the four inner planets (Mercury, Venus, Earth, and Mars) are made mostly of rock, while the four large outer planets (Jupiter, Saturn, Uranus, and Neptune) contain huge amounts of hydrogen and helium gas.

It was back to the drawing board, or more accurately, back to reconsider old ideas in a new light. The same efforts at calculation that led scientists to conclude that the near-collision idea would not work also helped them realize that Laplace's specific mechanism might not be the only way to form planets from a collapsing gas cloud. As they worked out the details anew, scientists soon found that the Kant-Laplace idea *could* explain nearly all the observed characteristics of our solar system. The idea returned to favor.

In science, it is difficult if not impossible ever to prove an idea true beyond all doubt. Nevertheless, it now seems a near certainty that our solar system did indeed form from the gravitational collapse of an interstellar gas cloud. Like any idea in science, this one has gained support because of evidence. In this case, the evidence is so overwhelming that the idea has risen in status to become what scientists call a *theory*. Note that, by this scientific definition, a theory is very different from a guess or a hypothesis; it is an idea that has been carefully checked and tested and that has passed every test yet presented to it. As we'll discuss more later, this difference in the way scientists define theory from the way it tends to be defined in everyday language explains why things like stickers reading "it's only a theory" don't make any scientific sense.

Part of the support for our current theory of the solar system's birth lies in the fact that it explains so many characteristics of our own solar system. Perhaps more important, the theory makes predictions that have been borne out with recent observations. In particular, it predicts that other star systems should form similarly from clouds of interstellar gas and that planets should be common around other stars. Both predictions have been verified. Scientists using the Hubble Space Telescope and other observatories have photographed stars that are in the process of being born today. These stars are clearly forming from the gravitational collapse of gas clouds, and they are forming in just the way our theory predicts they should form, with the stars surrounded by spinning disks of material just like the disk in which we think the planets of our own solar system formed. While these observations prove only that other stars have the *potential* to have planets around them, recent discoveries of bona fide planets demonstrate that, in at least some cases, the potential becomes a reality.

As recently as 1995, we still did not know for certain whether planets like those that orbit our Sun existed around any other star. In the little over a decade since, discoveries of extrasolar planets—planets in other solar systems—have come so rapidly that we now know of far more planets outside our solar system than within it. So far, most of these new planets are closer in size to Jupiter than to Earth, but that is probably just an artifact of the remarkable technology required to find them. I'll discuss this technology in some depth in chapter 8, but for now I can put it to you like this: Detecting a planet the size of Jupiter in another star system is rather like detecting a marble in a haystack from a distance of thousands of miles away. It is truly astonishing that we can now do this successfully for many Jupiter-size planets, and perhaps not too surprising that we cannot yet do it for planets the size of Earth, which would be like pinheads in the same haystack. Scientists are rapidly improving their planet-detection capabilities, however, and a NASA mission called *Kepler,* scheduled for launch in 2009, ought to be capable of finding at least a few Earth-size planets. Thus, if all goes well, within the next 5 to 10 years we will have a definitive answer to the question of whether planets similar in size to Earth exist around other stars. I'd bet my shirt that the answer will be "yes."

At the same time that we've been learning that planetary systems ought to be common, we've also been learning much more about what makes planets tick. In the inner solar system, we now understand why Venus is *so* much hotter than Earth, despite the fact that, relatively speaking, it is only slightly closer to the Sun. We understand why the Moon is desolate, even though it is essentially at exactly the same distance as Earth from the Sun. We even think we understand why Mars is cold and dry today, but shows clear evidence of having had rivers and perhaps seas in the distant past. In the outer solar system, we now understand why the large outer planets have moons that in some cases (such as Europa) might have underground oceans. This general understanding of planetary science means that we can evaluate the different worlds of our solar system in terms or their potential suitability for life, even though we are not yet capable of making definitive searches for life. The preliminary indications are promising—while we don't expect to find anything large or complex, at least a few other worlds in our own solar system seem good candidates for simple or microscopic life.

If you put all these ideas together, the planetary context for the search for life beyond Earth boils down to these three facts: First, there is at least some possibility that other worlds in our own solar system are capable of harboring life, although it would probably be very primitive life. Second, it is virtually inevitable that planets similar to those in our solar system exist in other star systems, making primitive life equally likely in those systems. Third, while we do not yet know it for sure, it seems likely that planets very much like Earth exist in many other star systems, opening up the possibility that they could harbor abundant and complex life—and perhaps even beings curious about whether life exists beyond their own world.

THE ASTRONOMICAL CONTEXT

The planetary context tells us that it is reasonable to imagine planets with life around other stars. But if we really want to understand just *how* reasonable it is, we need to turn to astronomy. Perhaps my background as an astronomer makes me biased, but it is the astronomical context that I find the most amazing of all.

The night sky may seem crowded with stars, but even under the best of conditions, you can see no more than a few thousand stars with your naked eye. If you want to understand the real meaning of the word "astronomical," you need to think about what lies beyond the naked eye limit.

I like to think about our place in the universe by considering what you might call our "cosmic address" (figure 1.1). We live on a planet, Earth, that is the third planet out from the star that we call the Sun. Our Sun, in turn, is one of a vast collection of stars that make up what we call the *Milky Way Galaxy*. Our galaxy travels through the universe along with about 40 other galaxies that, together, make up what astronomers call the *Local Group* of galaxies. Most other galaxies also reside in groups, which are called *clusters* when they have hundreds or thousands rather than just dozens of member galaxies. Groups and clusters are also grouped together, making what astronomers call *superclusters* of galaxies. Together, all the superclusters and all the spaces between them make up what we call our *universe*.

In terms of possibilities for life in the universe, the first thing to understand is that the universe is big, really BIG. I'll talk more about the scale of the universe in chapter 3, but for now let's just think about the number of stars and planets, starting with our own Milky Way Galaxy.

We do not know the precise number of stars in our galaxy, but it is at least 100 billion and perhaps one trillion or more. Are you wondering why

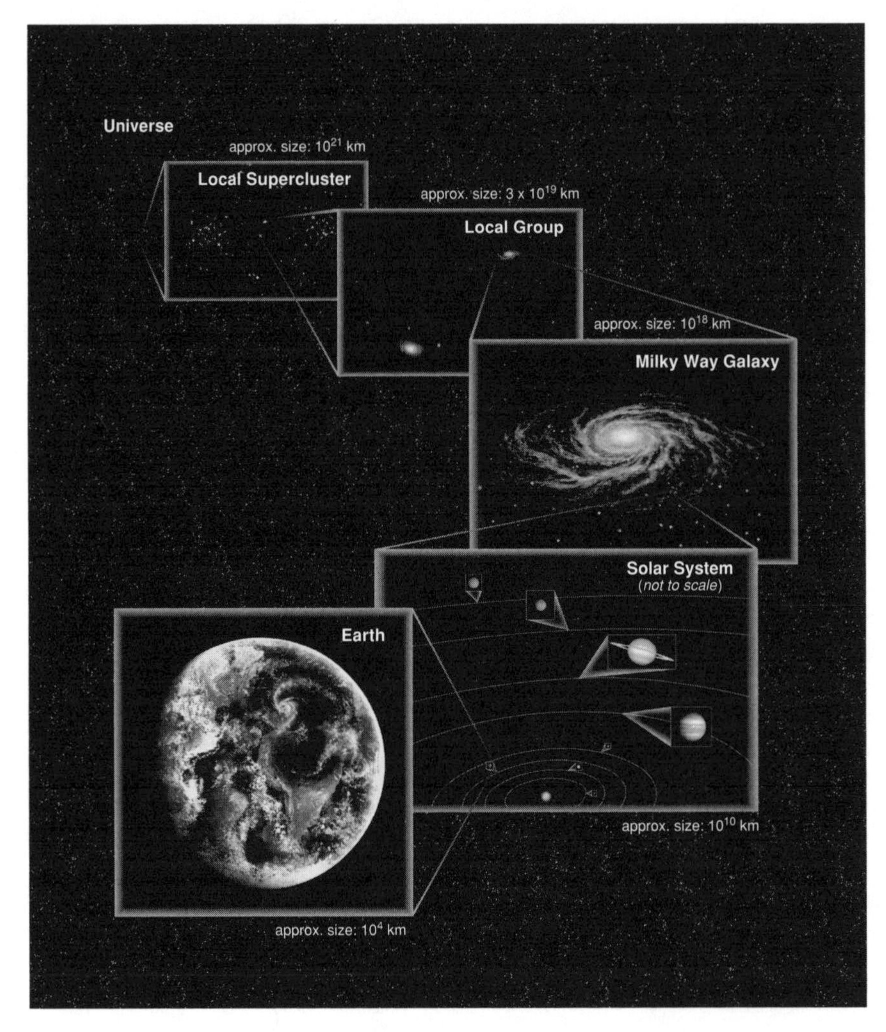

Figure 1.1. Our cosmic address. (Illustration courtesy of Addison Wesley, an imprint of Pearson Education)

we don't know the exact number? Imagine that you are having difficulty falling asleep tonight, perhaps because you are contemplating the possibilities of life beyond Earth. Instead of counting sheep, you decide to count stars. Let's be conservative, and suppose that our galaxy has only our minimum number of 100 billion stars. How long would it take you to count them? If you could count them at a rate of one per second, then it would obviously take you 100 billion seconds.[3] But how long is that? You can get the answer quite easily by dividing 100 billion seconds by 60 seconds per minute, 60 minutes per hour, 24 hours per day, and 365 days per year. If you do this calculation, you'll find that 100 billion seconds is nearly 3,200 years. In other words, you would need thousands of years just to *count* the stars in the Milky Way Galaxy, let alone to study them or search their planets for signs of life. And this assumes you never take a break—no sleeping, no eating, and absolutely no dying!

Now, take a look at the photo in color plate 1, which was made with 11 days of exposure time by the Hubble Space Telescope. To understand what you are seeing in this photo, imagine holding a grain of sand at arm's length against the sky; everything you see in this photograph would fit within the field of view directly behind that grain of sand. Almost every blob and dot that you see in the photo is an entire galaxy—each with so many stars that it would take thousands of years just to count them. Try to imagine the total number of stars located in this sand-grain-size piece of the sky, and then try to imagine the total number of stars in all directions around the entire sky. In truth, it's unimaginable, but I'll give you something that you can at least grasp onto: The total number of stars in the sky is roughly the same as the total number of grains of sand on *all* Earth's beaches, put together.

With as many stars as grains of sand on all Earth's beaches, it might seem almost impossible to believe that ours could be the only star orbited by a planet with life and a civilization. But numbers alone cannot tell the whole story. After all, if our solar system is very different from others—as would have been the case if, for example, the near-collision idea for the birth of the planets had turned out to be correct—then planets and life elsewhere might be quite unlikely. Since we do not yet have the ability to detect Earth-like planets around even nearby stars, we have no direct data from which we can decide whether such planets are common. However, everything we have learned about the universe since the Copernican revolution all points in the same direction: While we do not yet have proof of the existence of other planets like ours, we should expect them to be fairly common.

[3] When I give this question to children, they invariably object that they can count faster than that. However, while most kids can indeed count from 1 to 10 in less than 10 seconds, I like to point out that it's much more difficult to maintain a pace of one per second when you get to, say, "sixty-two billion, four hundred seventy-nine million, three hundred eighty-one thousand, five hundred forty-four" (and can you even remember what comes next?).

I say this because the central lesson of the Copernican revolution and nearly everything we have learned since has been that we are *not* central, after all. We are not the center of our solar system. Our Sun is not the center of the Milky Way Galaxy. Our galaxy is not the center of the Local Group. The Local Group is not the center of the Local Supercluster. The Local Supercluster is not the center of the universe; indeed, as we understand it today, the universe does not even *have* a center. Our place in the universe is completely ordinary, which makes it reasonable to think our planet is quite ordinary as well.

Could it be that, despite our ordinary location, there is something unusual about right here? Observations say no. We can measure the chemical compositions of distant stars, gas clouds, and galaxies by studying their spectra. The results tell us that the composition of our Sun and solar system are, like our location, ordinary. Spectra also tell us about the physical laws operating in distant objects; for example, if the laws of chemistry in distant stars were different from those on Earth, we'd be able to tell because the spectra of chemical elements in those stars would be different from the spectra of the same elements on Earth. But they are not, demonstrating that the same laws of nature act throughout the universe.

Our understanding of the origin of chemical elements gives us further reason to think that other stars should be like our Sun, making it possible for them to have planets like Earth. Observations show that chemical content of the universe consists almost entirely of just two elements: hydrogen and helium. These two lightest and simplest of the chemical elements make up at least 98 percent of the matter found in all stars and all gas clouds in space.[4] All the rest of the elements, from the carbon and oxygen that make up a large proportion of our bodies to the gold and silver that we wear as jewelry, make up no more than 2 percent of the overall chemical content of the universe. Moreover, we find that older stars have even smaller proportions than younger stars of elements besides hydrogen and helium, suggesting that the heavier elements have somehow been manufactured through time. I won't go into the details here, but we now think we know how: They were manufactured by nuclear fusion in

[4] When I speak of the chemical composition of the universe, I mean the "ordinary" matter made of atoms. As some readers may know, we now have reason to think that most of the mass of the universe consists of so-called "dark matter," which is presumably not chemical in nature. But this matter is not found in planets or stars, and thus should have little bearing on the search for life.

stars. In other words, we now think the universe was born containing only hydrogen and helium, and the rest of the elements have been made by stars. This idea implies that the same elements should be found in the same proportions everywhere, because the basic nature of stars is the same everywhere. It also implies that almost every atom in our bodies and in our planet Earth (except for the hydrogen) was made inside a star that lived and died before our Sun was born. As Carl Sagan was fond of saying, we are "star stuff."

Given that we live in an ordinary location in a solar system with ordinary composition and that the same laws act in all the other ordinary locations, is there anything else that could make our situation unusual? Some people point to time, but there's nothing special about the present, either: According to current understanding, our solar system was born about four and a half billion years ago, at a time when the universe as a whole was already nearly 10 billion years old. In other words, most of the stars in the universe are older than our own Sun, so even if life needs billions of years to arise and evolve into intelligence, plenty of stars with plenty of planets should have had plenty of time.

The last refuge of those who want to believe that our circumstances are unique is to imagine that it is some combination of multiple factors that, together, makes planets like Earth extremely rare. Proponents of this "rare Earth" hypothesis make some very interesting arguments, though as we'll see in chapter 8, there are also seemingly good counterarguments to each point they raise. Scientifically speaking, we simply do not yet have enough data to decide whether the rare Earth arguments have merit. But philosophically, and to remove the suspense, I'll tell you where I stand right now: For thousands of years, people have used every argument at their disposal to make the case that we humans somehow hold a central or special place in our universe. And every time that data have allowed us to evaluate one of those arguments in detail, the arguments have turned out to be flawed, removing us from our central place. I don't think the fate of the new "rare Earth" arguments will be any different from the fate of Aristotle's arguments about why other worlds could not exist at all.

So now you know how this chapter got its title: Like the number of grains of sand on all the beaches on Earth, our universe is filled with worlds that are truly beyond imagination. Neither I nor anyone else can yet prove that even a single one of those worlds harbors even the most primitive single-celled organisms, but it sure seems worth looking.

THE BIOLOGICAL CONTEXT

Together, the planetary and astronomical contexts tell us that we should expect to find lots of planets that are capable of harboring life. But the potential to have life and actually having life are not the same thing. Could it be that, even under perfect conditions, biology is extremely rare?

Until just a few decades ago, we did not even know where to start in addressing this question. The theory of evolution told us how life gradually changed through time, but by itself it gave no clue as to how life got started in the first place. The existence of life, with all of its biochemical complexity, remained beyond scientific understanding.

We still do not know how life on Earth got started, and it's possible that we never will. Nevertheless, recent biological discoveries give us at least some reason to think that life could prove to be almost as common as worlds capable of harboring it. Three lines of evidence point us in this direction.

First, laboratory experiments have demonstrated that chemical constituents found on the early Earth would have combined readily into more complex organic (carbon-based) molecules, including virtually all the building blocks of life (such as amino acids, nucleic acids, sugars, and lipids). Indeed, scientists have found organic molecules in meteorites and, through spectroscopy, in clouds of gas between the stars. The fact that organic molecules form even under the extreme conditions of space suggests that they form quite readily. In that case, the building blocks of life should be present on many worlds.

Of course, the mere presence of organic molecules does not necessarily mean that life will arise, but the history of life on Earth gives us some reason to think that it will. The relevant evidence comes from geological studies of the early Earth which, as we'll discuss further in chapter 5, tell us that life on Earth arose almost as early as it possibly could have after the Earth's formation. What does this early arrival of life on Earth prove? Absolutely nothing, because you cannot draw general conclusions from the single example of Earth. Nevertheless, it is at least suggestive of the idea that it's fairly easy for a planet to go from simply having organic material to actually having life. If the transition from organic chemistry to biology were difficult, we might expect that it would have required much more time. While we cannot say anything definitively, the early origin of life on Earth makes it reasonable to think that life would emerge just as quickly on other worlds with similar conditions.

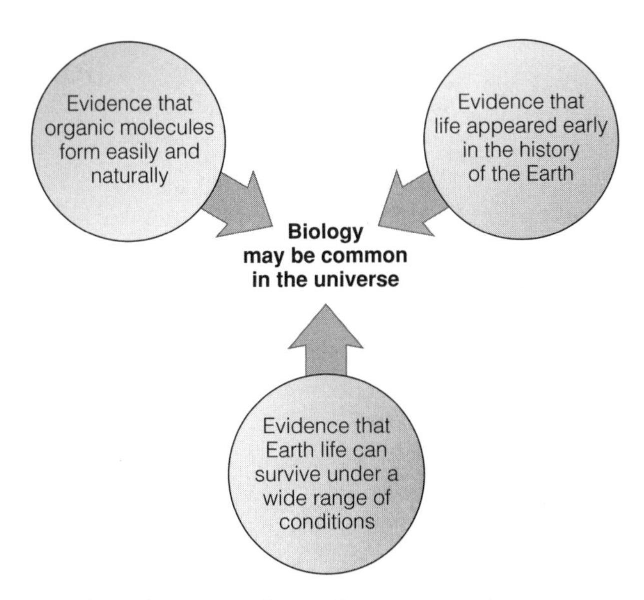

Figure 1.2. Three lines of evidence that give us at least some reason to think that biology may be common in the universe. (Illustration courtesy of Addison Wesley, an imprint of Pearson Education.)

If life really can emerge easily under the right conditions, the only remaining question is the prevalence of those "right" conditions. Here, too, recent discoveries give us reason to think that biology could be common. In particular, biologists have found that life can survive and prosper under a much wider range of conditions than was believed only a few decades ago. For example, we now know that life exists in extremely hot water near deep-sea volcanic vents, in the frigid conditions of Antarctica, and inside rocks buried a mile or more beneath the Earth's surface. If we were to export the strange organisms that live in these extreme environments to other worlds in our solar system—perhaps to Mars or Europa—it seems possible that at least some of them would survive. This suggests that the range of "right" conditions for life may be quite broad, in which case it might be possible to find life even on planets that are quite different in character from Earth. Figure 1.2 summarizes the three lines of evidence that suggest life could be common.

Now, before we go any further, it's important to address a question that is probably on many of your minds: Where does God fit into this picture? The way I've described the possibility of getting life, it may sound like it requires nothing more than random interactions of atoms, much as the

Greek atomists might have claimed some 2,300 years ago. But if you've followed my words closely, you'll see that I've said no such thing. In essence, we are still in the same place as Kepler and Galileo after they confirmed the Copernican idea: We know that Aristotle was wrong, but that doesn't necessarily mean the atomists were right. As far as current scientific evidence goes, we have no means of distinguishing whether we are a random accident in a universe without purpose or the pinnacle of creation in a miraculous process that God has directed from start to finish. So if a scientist tries to tell you that there's no room for God in our present understanding of life and evolution, he's just plain wrong: We may not have any scientific evidence of a role for God, but neither do we have any scientific evidence against it.

Of course, the same idea should also hold on the other side. The Bible is a complex and beautiful book that different people can interpret quite differently, even while believing that it is the word of God. Pope John Paul II, for example, believed in the literal truth of the Bible yet saw no contradiction between that truth and the scientific theory of evolution. If someone tries to tell you that science and evolution contradict the Bible, you can be quite certain that they are expressing their personal interpretation of God's words, not the actual words themselves. You can be a good Christian—or a good Jew, good Muslim, good Buddhist, or anything else—and a good scientist at the same time.

Indeed, the lack of conflict between science and religion seems to me so self-evident that I'm flabbergasted at the fact that not everyone else sees it the same way. Can't everyone just calm down, and realize that science and religion do not pose threats to one another? I say these things not just because I enjoy getting up on my soapbox (I admit it), but because I don't want anyone to miss out on the human joy of science. I am a scientist because I find the process of discovery to be inherently exciting, and I'm a writer because I want to share that excitement with others. I've chosen to write about the scientific search for life in the universe because, in my opinion, it is a topic brimming with more excitement than any other. It may not qualify as the greatest story ever told, but it's a darn good one, and if and when we find other life or other civilizations, I believe that it will cause a revolution in the way we think about ourselves that will be every bit as profound as the revolution that occurred some 400 years ago when we learned that Earth moves. I'd like to think that everyone, regardless of culture or religion, can be a part of this ongoing story of discovery. So perhaps I'm too naive . . . but, at least, I hope that those of you with deeply religious beliefs will not feel threatened by reading the rest of this book.

BEYOND UFOS

I've briefly addressed religion, so now there's one more group of people I need to address before we go on: the roughly half of the public who, according to polls, believe we are already being visited by UFOs. Rest easy, because I will not tell you that you are wrong.

How could I? I've spent the entire chapter explaining why, according to current scientific understanding, it is eminently reasonable to think that life could be quite common on worlds that number beyond imagination. And while we haven't yet discussed the scientific issues that differentiate getting intelligent life and civilizations from just getting life of some kind (that will come in chapters 5 and 9), sheer numbers suggest that if life is very common, civilizations ought to be at least somewhat common. Moreover, if civilizations are common, the age of the universe ought to ensure that many of them have had time to advance far beyond us technologically, in which case they might well have the ability to travel from their home worlds to here. As I see it, it would not be at all surprising if aliens really are visiting Earth.

Still, I am personally very skeptical of any and all the claims I've ever heard of UFOs and other alien visitation to Earth. This might sound strange: How can I say that alien visitation is likely and then, in nearly the same breath, doubt the reports of visits? My answer is twofold. First, there's the issue of evidence. In science, we can't accept an idea just because it's reasonable; we need verifiable evidence, and the evidence presented for UFOs just doesn't measure up to scientific standards. Second, once you understand the technology that aliens must have if they really are visiting us, you'll see that most of the claims that people make about the supposed visits don't make any sense. But don't just take my word for these things now; read on, and in the next two chapters I'll explain these ideas and their remarkable consequences.

2

WHAT MAKES IT SCIENCE?

> All our science, measured against reality, is primitive and childlike—and yet it is the most precious thing we have.
>
> —*Albert Einstein*

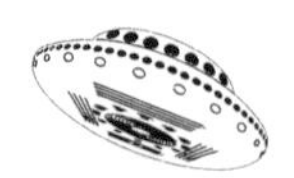

I'd always wanted to see a real UFO—something in the sky that I could not explain and that would therefore qualify as an *unidentified* flying object. Then, even without proof, I could at least hope that I'd seen an alien spacecraft. For most of my life, it never happened. Sure, I saw lots of strange things in the sky. But with a little thought, I'd soon conclude that I'd only seen a distant airplane or a rocket trail or the planet Venus seeming to dart about as clouds passed in front of it. Ironically, I finally saw my first UFO just a few weeks after I started working on this book.

I was outside with my then 6-year-old son, Grant, watching the stars in the predawn sky. Venus was shining brightly in the east, which made me do a double take when I suddenly saw another object shining just as brightly in the west. Over the next few seconds, the object grew brighter and brighter until it was by far the brightest object in the sky. I called to Grant to look over at it. "Wow!" he said. Then, as quickly as it had brightened, it faded away. To my eyes it merely disappeared. But Grant, who as a child has much better night vision, said it darted off to the right as it vanished from view. The entire episode lasted no more than about 10 seconds.

No airplane could have moved in that way, nor could it have been a satellite or rocket trail. It wasn't a planet, and it wasn't a cloud. In fact, my first thought as I watched it brighten was that I was witnessing the explosion of a distant star—a nova or a supernova. But its rapid disappearance ruled out this idea, because stellar explosions take days or weeks to fade from view. So what was it? Had I finally witnessed an alien spacecraft flying in for a quick glimpse of my town?

Possibly, but I also came up with an alternative explanation. I love my sleep almost as much as I love the stars, and we were outside at 4 AM only because it was the night of the annual Perseid meteor shower. By the time we saw the strange light, Grant and I had already seen a couple dozen meteors streaking across the sky. Could our strange light simply have been another meteor?

Meteors are created when pebble-size pieces of dust from space burn up high in our atmosphere. Particles of space dust typically plunge into the atmosphere at a speed in excess of 30,000 miles per hour. This high speed generates intense friction, making the air around the particle so hot that it glows. In other words, the flash of a meteor is the glow of hot air surrounding a high-speed particle, rather than the particle itself. The flash ends when the particle has fully disintegrated. Most of the dust particles that crash into Earth were shed by comets that passed near Earth's orbit. We get annual meteor showers because our planet crosses the same trails of comet dust at the same time each year. The Perseid meteors get their name because the geometry of the meteor shower makes the meteors appear to emanate from the constellation Perseus as they burn up in the atmosphere.

The trouble with thinking of my light in the sky as a meteor is that it didn't act like a Perseid meteor should. It did not streak across the sky, and it did not appear to come from the direction of Perseus. In fact, because it appeared to stay nearly stationary as it brightened and faded, it could have been a meteor only if had been coming almost straight toward us, so that it would appear motionless but brighter as it came closer. Even then, I still don't have a good explanation for the sudden movement that Grant saw at the end. Perhaps what he saw was a secondary flash as a fragment of the dust particle flew off in one direction. Or perhaps he saw an illusion created by the movement of his own head.

The bottom line is that I cannot conclusively identify the light I saw in the sky as a meteor or as anything else, which means I can truly claim to have seen an unidentified flying object. However, I cannot automatically conclude that my UFO was an alien spacecraft. It might indeed have been evidence of alien visitation, but it might also have been a rather unusual meteor. The heart of science lies in the way we choose among such competing explanations.

If I let my wishes get the best of me, I would choose the explanation of alien visitors. After all, I really want to believe that the universe is full of life and that we'll someday make contact with other civilizations. If I could just accept the idea that I experienced such contact on the night when I saw the light in the sky, then I could join the legions of people who believe with all their hearts that aliens are here among us.

But I'm either blessed or cursed with a scientific mind, and I'd therefore bet about 25 million to one that my UFO was actually an unusual meteor rather than an alien spacecraft. Why? Because every day, about 25 million pieces of space dust enter the atmosphere and burn up as meteors somewhere in Earth's sky. It seems far more reasonable to think that I saw an odd one among those 25 million rather than something as extraordinary as beings from another world. I'm reminded of a dictum from the great Carl Sagan: "Extraordinary claims require extraordinary evidence."

I tell this story not to discredit other UFO sightings but rather to emphasize what I consider to be the most basic difference between science and beliefs. Science is supposed to be based on verifiable evidence, while beliefs are matters of faith or opinion. I could believe with all my heart that I really did see an alien spacecraft, but if you don't believe me, there's nothing that either of us can do to convince the other. For an idea to be science, it has to be something that we *can* come to agreement on, at least in principle, by comparing notes on evidence that we both can study.

The idea that science is a way of helping people come to agreement may seem surprising in light of the cultural wars we often read about in the news, but it explains why science has been so successful in advancing human knowledge. Think back to the debate between Aristotle and the atomists over the question of whether there could be worlds beyond Earth. For nearly 2,000 years, this ongoing debate went essentially nowhere, because there was no way for the two sides to come to agreement on any of the issues involved. But as soon as we had solid, verifiable evidence showing that Earth is a planet going around the Sun, we knew that Aristotle's position had been wrong. To be fair, while Aristotle turned out to be fundamentally incorrect in many of his beliefs about physics and astronomy, he was actually quite a good observer and made many important discoveries in other subject areas. In biology, for example, he correctly described numerous relationships between species. Where he erred, he did so because he had nothing solid to go on. If Aristotle could have returned to life in the mid-1600s and examined the overwhelming evidence demonstrating that Earth is not the center of the universe, I think he would have been quite convinced. The evidence certainly convinced the scientific community, and the agreement on this point led people to ask new questions, such as what holds Earth in its orbit as it goes around the Sun. The quest to answer these new questions ignited the scientific and technological revolution that has made our civilization what it is today.

OK, you may ask, but if science is supposed to help people come to agreement, why does it so often seem to do the opposite? Why, for example, do some religious people think that science is out to destroy their faith, and why do so many UFO believers think that science is trying to hide the truth from them? Honestly, I don't really know, but my suspicion is that those who feel threatened by science don't really understand it. If they did, they'd realize that science is indeed a tool for bringing people together with common understandings. I doubt that anyone could find fault with that, no matter what their personal views of God.

So as I step back down off my soapbox, you can probably see where we're going next. If I'm going to achieve this book's stated goal of helping you understand what science really tells us about extraterrestrial life, we need to be very clear about what science is and what it is not. Otherwise, we'll be stuck like Aristotle and the atomists in endless debate over things like my UFO in the night sky, talking and talking but never actually learning anything.

THE ANCIENT ROOTS OF SCIENCE

Just as we can understand a fellow human being better by knowing what she experienced in childhood, we can understand science better if we understand how it grew up. Science grew up primarily through attempts to understand the motions of the Sun, Moon, planets, and stars in the sky.

Why is it that astronomy was so important to ancient people? The first answer was practicality. In the days before mechanical and electronic devices, the only clocks and calendars were in the sky. If you wanted to know the time of day or the time of year—clearly critical information for agrarian societies—you had to know how to read them from careful observations of the Sun's position in the sky. Around the world, you can still see many amazing structures constructed largely for the purpose of telling the time or date by the Sun; famous examples include Stonehenge, Egyptian obelisks, and the Native American Sun Dagger in New Mexico. The Moon was only slightly less important. Many civilizations grew up along coastlines, so keeping track of the Moon enabled them to predict and work with the tides.

The practical importance of marking the motions of the heavens did not automatically mean that people needed to understand *why* the movements occurred. After all, you can use a watch without knowing what's going on inside of it. But just as many kids like to take watches apart to see what makes them tick, ancient people were curious about the clockwork of the sky.

Early on, in what we might call mythological times, people tended to attribute what they saw in the sky to the supernatural. If you imagined the Sun or the planets as gods, it was easy to "explain" their motions as the prerogative of those gods. Science began as people tried to move beyond the supernatural, instead trying to come up with physical mechanisms by which they could not only explain what they saw in the sky but also predict what they would see in the future. Because most ancient cultures left relatively few written records—and those few were more likely to be about politics or religion than about the search for physical explanations—we really do not know how many people in how many different cultures might have been early practitioners of science. What we do know is that this type of science was under way in Greece by about 500 B.C., and that we can trace a nearly straight line from the ancient Greek meditations to the methods of modern science.

As we discussed briefly in chapter 1, the Greeks generally assumed that Earth lay unmoving at the center of the universe, a very natural idea given that our world feels quite stationary and the sky appears to circle around us. But if they were actually going to predict the future positions of the Sun, Moon, and planets in the sky, the Greeks needed much more than just this idea. They needed a physical *model* of the universe, one that would allow them to calculate future positions with the aid of mathematics.

The first step in creating such a model is to search for patterns of motion that the model must explain. The motion of the stars was very easy: The stars stay fixed in the same constellations from night to night and year to year, and all seem simply to circle around our world once each day. Thus, to explain the motions of the stars, the Greeks envisioned a great, rotating celestial sphere surrounding our central world, with the stars arranged on the great sphere in the patterns of the constellations.

The Sun follows an only slightly more complex pattern of motion in our sky: It circles daily around us much like any star, but over the course of the year it gradually moves through the constellations along the path that we call the *ecliptic*. The Greek philosophers could explain this motion by imagining that the Sun turned around Earth on its own sphere, with the turning rate tuned so that from our viewpoint it reproduced the Sun's annual motion along the ecliptic. A third sphere took care of the Moon, which also moves steadily through the constellations, though not precisely on the same path from one month to the next.

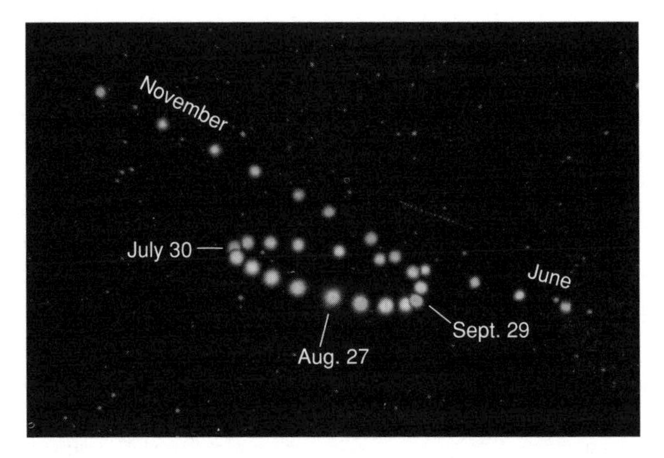

Figure 2.1. This composite of 29 photographs, each taken at five- to eight-day intervals, shows Mars between early June and late November 2003. Notice the period of "backward" motion in the middle of the loop. The white dots in a line just right of center are the planet Uranus, which by coincidence was in the same part of the sky. Photo by Tunc Tezel.

The real difficulties for the Greek model came with the planets. Unlike the Sun and Moon, the planets do not move steadily through the constellations. The Sun moves along the ecliptic at a rate of just under 1 degree per day, which is why it takes just a few more than 360 days to circle all the way around. The Moon moves through the constellations somewhat faster—about 12 degrees per day, which is enough that you can notice this motion even on a single night by comparing the Moon's position relative to a bright star early in the evening to its position a few hours later. The planets, in contrast, seem to move among the stars in a very erratic way. Sometimes they move relatively fast from one night to the next, other times more slowly. Most strangely of all, they sometimes reverse course entirely, moving "backward" relative to the stars for a few weeks or months—a phenomenon known as *apparent retrograde motion*. For example, the composite photo in figure 2.1 shows Mars over a period of about six months; notice its retrograde loop in the middle, during which it turned around and moved "backward" compared to its normal direction of motion through the constellations.

This complex motion could not be explained just by adding another sphere for each planet, unless you were to allow the sphere's rate and direction of rotation to vary over time. But the Greeks did not allow such variations, in part because such arbitrary variations still wouldn't have offered a set of rules by which to predict future planetary positions, but more importantly because it would have violated a central tenet of Greek thought. In a doctrine enunciated most clearly by Plato (428–348 B.C.), the Greeks held that the heavens must be "perfect," which they took to mean that heavenly objects must move in "perfect" circles with perfectly constant speeds. This doctrine was so deeply ingrained in Greek thought that, as far as we know, they never seriously considered dumping it, no matter how much it seemed to disagree with observations. And why did they hold this doctrine so dear? We really don't know; they just did. It certainly wasn't backed by any actual evidence. It was just something they believed.

In any event, faced with the reality that the planets sometimes move backward relative to the stars, the Greeks faced essentially two choices for how they could go about trying to explain this phenomenon. Behind Door #1 (so to speak) lay the choice that we now know to be the truth: The planets don't really ever go backward, they just seem to as we pass by them in our orbit of the Sun. You can see how this works with the simple demonstration shown figure 2.2. Have a friend walk in a circle to represent Mars's orbit while you walk in a circle to represent Earth's orbit; be sure you walk faster than your friend, since inner planets orbit the Sun faster than outer planets. If you watch your friend's position against objects in the background, you'll see that your friend *seems* to go backward as you "lap" her in your orbit, even though she never really reverses course. As the second illustration shows, the same idea explains why we see the real Mars sometimes move backward relative to the stars. Alternatively, the Greeks could choose the explanation behind Door #2, in which they could hold to their notion of spheres surrounding Earth by attempting to come up with an extraordinarily convoluted and complex model of spheres turning inside of other spheres, with each sphere in some different rotational orientation, with the ultimate hope of making the whole combination lead to something that would predict planetary positions at least moderately well.

With our modern-day hindsight, it might seem strange to think that anyone would choose Door #2 over Door #1, but with a few notable exceptions—such as Aristarchus, whom we encountered in chapter 1—that's exactly what the Greeks did. Why did they choose a complex and convoluted explanation when a far simpler one was available? In part, it's because the correct answer would have forced them to throw out the idea of Earth as the center of the universe, and many of them probably thought that to be a far too radical solution. However, many Greek philosophers apparently gave

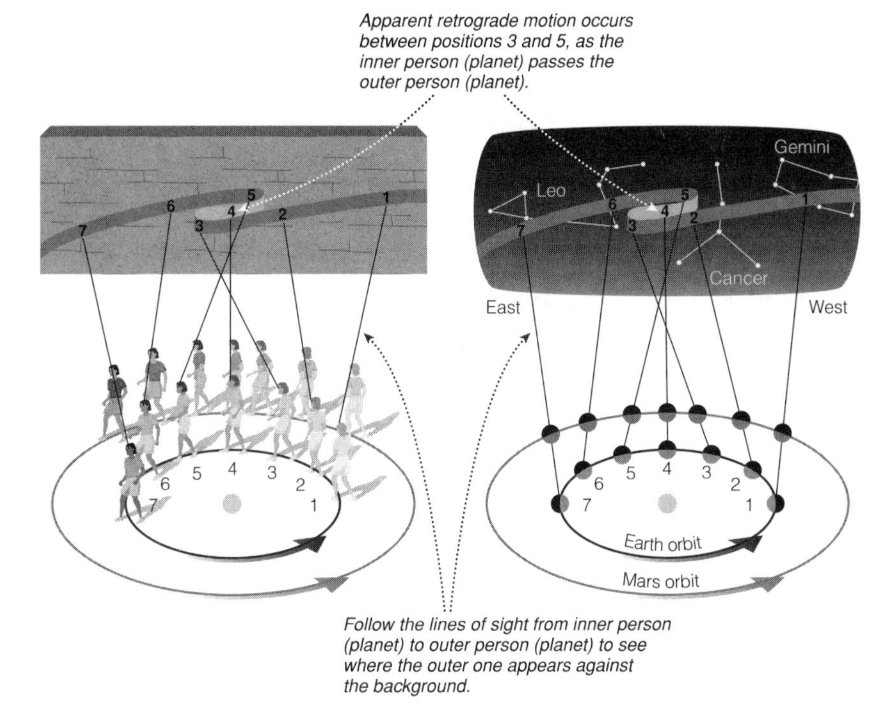

Figure 2.2. The real explanation for the fact that planets sometimes move "backward" relative to the stars: The demonstration on the left shows the basic idea, and the diagram on the right shows how it applies to Mars. (Illustration courtesy of Addison Wesley, an imprint of Pearson Education)

serious consideration to Aristarchus's suggestion, and actually rejected it on its merits as they were understood at the time.

In particular, Aristarchus's idea seemed inconsistent with observations of stellar positions in the sky. To understand why, imagine what would happen if you placed the Sun rather than Earth at the center of the celestial sphere, with Earth orbiting the Sun some distance away. In that case, Earth would be closer to different portions of the celestial sphere at different times of year. When we were closer to a particular part of the sphere, the stars on that part of the sphere would appear more widely separated than they would when we were farther from that part of the sphere, just as the spacing between the two headlights on a car looks greater when you are closer to the car. This would create annual shifts in the separations of stars—but the Greeks observed no such shifts. They knew that there were only two possible ways to account for the lack of an observed shift: Either Earth was at the center of the universe, or the stars were so far away as to make the shift undetectable by eye. To most Greeks, it seemed unreasonable to imagine that the stars could be *that* far away, which therefore left them with the conclusion that Earth must hold a central place.

Significantly, this basic argument still holds when we allow for the reality that stars lie at different distances rather than all on the same sphere. As Earth orbits the Sun, we look at the stars from different positions in space at different times of year. Just as your finger will seem to shift back and forth against the background if you hold it arm's length and alternately blink your left and right eyes, nearby stars should seem to shift against the background of more distant stars as we look at them at different times of year from opposite sides of Earth's orbit. Although such shifts (called *stellar parallax*) are much too small to measure with the naked eye, they are easily detectable with modern telescopes and therefore represent concrete proof that Earth really does go around the Sun. In fact, precise measurement of these shifts provides us a direct way to measure the distances to stars; the method is essentially the same method of "triangulation" that construction workers use to measure distances here on the ground.

Unable to detect these stellar shifts and therefore having concluded that Earth must really be in the center of things, for several centuries the Greeks developed new and ever-more complex ways of getting their Earth-centered model to make predictions that agreed with reality. This long effort culminated with the work of the Greek astronomer Ptolemy (pronounced *TOL-e-mee;* c. A.D. 100–170), who published a detailed and mathematically precise treatise that could be used to predict the future positions of planets among the stars. The required calculations were both very complex and extraordinarily tedious; many centuries later, while supervising calculations based on the Ptolemaic model, the Spanish monarch Alphonso X (1221–1284) is said to have complained that "If I had been present at the creation, I would have recommended a simpler design for the universe." Nevertheless, Ptolemy's model worked remarkably well, as it generally allowed planetary positions to be predicted to an accuracy of a few degrees—roughly equivalent to the size of your hand viewed at arm's length against the sky. This was an astonishing achievement at the time, and may be even more impressive with modern hindsight, since we now know that Ptolemy got these good answers even though he started from the fundamentally wrong idea that Earth is the center of the universe. When Arabic scholars translated Ptolemy's book describing the model, around A.D. 800, they gave it the title *Almagest*, derived from words meaning "the greatest compilation."

The great success of Ptolemy's model represented both the best and the worst of ancient Greek science. On the positive side, the model gained acceptance because it made predictions that agreed reasonably well with reality, and insistence on such agreement remains at the heart of modern science today. On the negative side, the model was so convoluted that it's unlikely that anyone, including Ptolemy himself, thought that it actually represented the true nature of the cosmos. Indeed, the model was not even fully self-consistent, as different mathematical tricks needed to be used to calculate the positions of different planets. Today, these negatives would weigh so heavily against any scientific idea that people would go immediately back to the drawing board in search of something that worked better. But in Ptolemy's time, these negatives were apparently acceptable, and it was another 1,500 years before they were revisited.

THE COPERNICAN REVOLUTION

The Greek ideas gained great influence in the ancient world, in large part because the Greeks proved to be as adept at politics and war as they were at philosophy. In about 330 B.C., Alexander the Great began a series of conquests that expanded the Greek Empire throughout the Middle East. Alexander was deeply interested in science and education, perhaps because he grew up with Aristotle as his personal tutor. Alexander established the city of Alexandria in Egypt, which soon became home to the greatest library the world had ever seen. The Library of Alexandria remained the world's preeminent center of research for some 700 years. At its peak, it may have held more than a half million books, all handwritten on papyrus scrolls. When the library was finally destroyed during a time of anti-intellectual fervor in the fifth century A.D., most of the ancient Greek writings were lost forever.

Much more would have been lost if not for the rise of a new center of intellectual achievement in Baghdad (in present-day Iraq). While European civilization fell into the Dark Ages, scholars of the new religion of Islam sought knowledge of mathematics and astronomy in hopes of better understanding the wisdom of Allah. The Islamic scholars—often working collaboratively with Christians and Jews—translated and thereby saved many of the remaining ancient Greek works. Building on what they learned from the Greek manuscripts, they went on to develop the mathematics of algebra as well as many new instruments and techniques for astronomical observation.

The Islamic world of the Middle Ages was in frequent contact with Hindu scholars from India, who in turn brought ideas and discoveries from China. Hence, the intellectual center in Baghdad achieved a synthesis of the surviving work of the ancient Greeks, the Indians, the Chinese, and the contributions of its own scholars. This accumulated knowledge spread throughout the Byzantine Empire (the eastern part of the former Roman Empire). When the Byzantine capital of Constantinople (modern-day Istanbul) fell in 1453, many Eastern scholars headed west to Europe, carrying with them the knowledge that helped ignite the European Renaissance.

The Renaissance brought a new spirit of inquiry, and technology helped fuel its spread. The most significant new technology was the printing press with movable type, invented by Johannes Gutenberg around 1450. Prior to its invention, books had to be laboriously copied by hand or printed from hand-carved pages of type. Indeed, books were so expensive and rare that few people had access to them, which is probably a major reason why most people of the time remained illiterate. The printing press changed all that. By 1500, some 9 million printed copies of some 30,000 works were in circulation. With books cheap and widely available, many more people learned to read. This had the effect of democratizing knowledge and naturally led to a much larger pool of scholars. The stage was set for a dramatic rethinking of our place in the universe, and of the principles of science as a means for advancing knowledge. The revolution began with Nicholas Copernicus (1473–1543).

The dramatic story of the Copernican revolution has been recounted many times; if you are interested in details, I encourage you to read (or watch) Carl Sagan's *Cosmos*, which is where I first learned of many of these details myself. Here, I want to focus only on how the Copernican revolution helped shape the nature of modern science.

Copernicus has his name attached to the revolution because he started people thinking about whether they should switch to his new Sun-centered model or stick with Ptolemy's old Earth-centered model. However, while his new model generated intense interest in the scholarly community, it actually won very few converts in the decades after its publication, and this failure came about for a very good reason: Despite having put Earth in its correct place, Copernicus's model did no better than Ptolemy's at predicting planetary positions.

Why didn't it work better? Blame it on Plato. In removing Earth from its central position, Copernicus willingly overthrew thousands of years of tradition and suggested an idea that would radically alter our view of our place in the universe. But he was not willing to overturn Plato's ancient dictum that heavenly motion must be in perfect circles. Because planets do *not* really orbit the Sun in perfect circles, Copernicus could not get his model to work any better than Ptolemy's, even when he added his own set of rather unrealistic orbital complexities.

Still, with things like its much more natural explanation for retrograde motion of the planets, the Sun-centered model offered an aesthetic attraction that many other scientists could not resist. Instead of ignoring Copernicus, they sought ways to make his model work better. The key players in this effort were Tycho Brahe (1546–1601) and his one-time apprentice, Johannes Kepler (1571–1630). Tycho, recognizing the importance of quality data against which any model could be checked, spent some three decades carefully recording what were by far the most accurate observations of planetary positions that had ever been made. Working before the invention of the telescope, he built large, naked-eye observatories that worked much like giant protractors, and he used them to measure planetary positions in the sky accurate to within 1 minute of arc—equivalent to less than the thickness of a fingernail held at arm's length. Kepler inherited these data after Tycho's death in 1601, and then set about trying to come up with a model of planetary motion that could match Tycho's observations.

Copernicus was a revolutionary, Tycho collected the key data, and later figures like Galileo and Newton did the work that sealed the case for the Copernican revolution. But for my money, it is Kepler to whom we most owe the birth of modern civilization, because he did something that no one else had been willing to do in the preceding 2,000 years: He trusted the data more than he trusted his own deeply held beliefs.

Kepler was a devout Christian and believed that understanding the geometry of the heavens would bring him closer to God. Like Copernicus, he believed in Plato's dictum about circular motion in the heavens, so he worked diligently to match circular orbits to Tycho's data. After years of effort, he found a set of circular orbits that matched Tycho's observations quite well. Even in the worst cases, which were for the planet Mars, Kepler's predicted positions differed from Tycho's observations by only about 8 arcminutes—meaning that his model predicted a position that differed from Tycho's written position by an amount barely one-fourth the angular size of the full moon. Ask yourself: What would you have done in Kepler's place, having spent years developing a model that was *that close* to perfection? Would you have said, "Well, Tycho must have made a mistake when he recorded those few observations that don't match my work"? Or would you have trusted the data, thrown out your years of effort, chucked your deep belief in perfect circles, and started all over again? It gives me goose bumps every time I really think about the fact that Kepler chose option 2. About this choice, it is worth reading the words of Kepler himself:

> *If I had believed that we could ignore these eight minutes [of arc], I would have patched up my hypothesis accordingly. But, since it was not permissible to ignore, those eight minutes pointed the road to a complete reformation in astronomy.*

Kepler abandoned perfect circles and began testing other orbital shapes. It again took him some years of work, but he finally hit upon the correct answer: Planetary orbits are not circles, but rather are the special types of ovals known as *ellipses*. Using his talents at mathematics, he worked out the mathematical details of the elliptical orbits, which we now describe as *Kepler's laws of planetary motion*. With these laws, anyone could predict the past, present, or future positions of any of the planets known at the time. Kepler's model not only produced a perfect match to Tycho's data, but its predictions of future planetary positions were also a perfect match to what was eventually observed. I'm no historian myself, but as I understand it from Harvard historian Owen Gingerich, one of the most crucial events occurred in 1631, a couple of decades after Kepler published his model. During that year, astronomers observed a relatively rare event called a *transit* of Mercury, when Mercury appears to pass directly across the face of the Sun. The transit occurred precisely as Kepler's laws predicted it would. Neither Ptolemy's model nor Copernicus's model nor any other model that anyone came up with could claim the same success.

It's important to realize that the failure of other models did not necessarily mean that Kepler's model was right, and even the great success of Kepler's laws did not *prove* they are true. Indeed, while there are many cases in the history of science where a model has been proven wrong, it is virtually impossible to prove a model right. The reason is that no matter how many successes a model may have, you can never be absolutely certain that it will still work in new cases. If, after Kepler's work, astronomers had discovered a new planet that did not obey Kepler's laws, they would have been forced to conclude that Kepler's laws did not always work and therefore would have either modified them or dumped them in favor of something else. That's just the way science works.

In fact, Kepler's laws are *not* perfect. Applied strictly, we know of many cases in which planets deviate from them in small but measurable ways. So why do we still accept that Kepler was right about planets orbiting the Sun along elliptical paths? The answer to this question is a key to understanding the difference between science today and science in ancient times, and hence to understanding why human knowledge is now advancing so rapidly: Back in ancient times, once a model (such as Ptolemy's) worked "good enough," people basically left it at that. But in modern science, we turn every answer into the next question.

From the moment that Kepler published his laws of planetary motion, other scientists asked questions about them. Some questioned whether they were consistent with other physical laws, since the idea of a moving Earth violated Aristotle's still-popular claims of natural motion. It took Galileo's work to prove that Aristotle had been wrong about physics, too, and thereby to seal the triumph of the Copernican idea. (Galileo's telescopic observations also played a major role, since these observations were consistent with Earth going around the Sun but could not be explained by the geocentric model.) Other scientists asked *why* Kepler's laws worked so well; after all, there was no known reason why orbits should be ellipses rather than circles or even squares. Scientists wrestled with this question for nearly 70 years before Sir Isaac Newton (1642–1727) came up with an answer. And in finding this answer, Newton not only discovered the more general laws of motion and the law of gravity, but he also had to invent the mathematics of calculus in order to prove that these new laws did indeed explain Kepler's laws.

Newton's laws did not stop the questioning either, even after they proved so successful that later scientists used them to *predict* the existence of the planet Neptune before it was actually discovered through a telescope. Message to those who believe their horoscopes: Astrology claims to be able to predict the future based on planetary positions among the stars; and, yet, for thousands of years, no astrologer ever realized that they were missing an undiscovered planet that is more than a dozen times as large as Earth. Astronomy found it, astrology didn't. That doesn't necessarily prove that astronomy will always be right, but it sure looks bad for the competition.

So while the astrologers just added Neptune to their horoscopes and went on like nothing else had changed, the astronomers keep questioning. And, by the late 1800s, they had indeed found something that didn't perfectly match the predictions of Newton's law of gravity. It was Mercury's orbit that wasn't quite obeying Newton, and it forced scientists to think again about what might be going on. The eventual result was Einstein's general theory of relativity, which gives essentially the same answers as Newton's theory for planets farther from the Sun, but a slightly different answer for close-in Mercury—an answer that matches the observations. In other words, at its core, Einstein's theory is a description of gravity, and it is the best description of gravity that we have because it works in every case that Newton's description worked and more. But scientists kept questioning, and today we know that even Einstein's theory cannot be the entire story, because it fails to explain what happens to gravity on the smallest, subatomic scales.

The quest to find an improvement on Einstein is one of the driving forces in physics today. Scientists have a lot of ideas about what this improvement might look like but, so far, no actual evidence with which to choose among the competing ideas. As a result, today we are in a position of knowing that a deeper understanding of gravity must be out there, but not knowing what it actually is. This type of unanswered question is what makes science so exciting, and it drives home the point I began with, that science is a way of helping people come to agreement. Today, many different scientists have many different ideas about what the new theory of gravity should be, but ultimately, when the evidence comes in, we'll be able to choose among the competing ideas and come to agreement about which ones must go and which ones are worth taking forward into the future.

HALLMARKS OF MODERN SCIENCE

We've discussed how the Copernican revolution gave rise to modern science, but we still haven't said exactly what science is. Indeed, you may have noticed that I've described science in several different ways already. In chapter 1, I said that science is a way of distinguishing possibilities from realities. In this chapter, I've said that science is a way of choosing among alternate explanations, and of getting people to agree. All these things are true, but they don't give us a clear way of deciding what qualifies as science and what does not. For that, we need a clearer definition of science.

Defining science is a surprisingly tall order. The word itself comes from the Latin *scientia,* meaning "knowledge," but not all knowledge is science. For example, you may know what music you like best, but your musical taste is not a result of scientific study. So what exactly is it that makes something science?

Scientists, historians, and philosophers have written hundreds of books and articles attempting to come up with a clear definition of science. Not everyone agrees on all the key points, which we can take as an illustration of the fact that semantics is not itself a science, since it does not offer us a clear way to come to agreement. Nevertheless, if you sift through all the history from the Greeks to the Copernican revolution and beyond, I believe that you'll find that everything that qualifies as science shares the following three characteristics, which I will refer to as the three hallmarks of science:[1]

- Modern science seeks explanations for observed phenomena that rely solely on natural causes.
- Science progresses through the creation and testing of models of nature that explain the observations as simply as possible.
- A scientific model must make testable predictions about natural phenomena that would force us to revise or abandon the model if the predictions do not agree with observations.

We can see each of these hallmarks in the story of the Copernican revolution. The first shows up in the way Tycho's careful observations of planetary motion motivated Kepler to come up with a better explanation for those motions. The second is evident in the way several competing models were compared and tested, most notably those of Ptolemy, Copernicus, and Kepler. We see the third in the fact that each model could make precise predictions about the future motions of the Sun, Moon, planets, and stars in our sky. Kepler's model gained acceptance because it worked, while the competing models lost favor because their predictions failed to match the observations.

These three hallmarks are so important that it's worth considering each of them in a little more detail. Let's start with the first, which happens to lie at the root of the current debate about whether "intelligent design" should be taught in science classes. Proponents of intelligent design claim that life is so intricate and complex that it could not have arisen naturally, and they therefore claim that life must have been deliberately designed by an intelligent Designer. Personally, I find their evidence of design far less than compelling, but that's really beside the point. The real question is whether their idea should qualify as a competing scientific model that could then be taught as an alternative to the theory of evolution. If you ac-

[1]These three hallmarks are not part of any generally accepted definition of science; rather, they are something that I have come up with in consultation with my textbook coauthors (especially Mark Voit, Megan Donahue, Nick Schneider, Seth Shostak, and Bruce Jakosky) and professors who use my textbooks. So far, we have received very positive feedback about using these hallmarks as a definition of science, but we are always looking to improve them as more people examine them.

cept the usual definition of science, then intelligent design clearly does not qualify, because it violates the first hallmark: Rather than seeking natural causes for life, intelligent design posits that life is the work of a supernatural Designer[2] who is beyond our scientific comprehension. That is why those who want to teach "ID" in science classes (such as the Kansas Board of Education in 2005) have attempted to redefine science so that it does not have to be solely about natural causes.

The trouble with these attempts to redefine the first hallmark is that they would render science pointless. As a simple analogy, consider the collapse of a bridge. If you choose to believe that the collapse was an act of God, you might well be right—but this belief won't help you design a better bridge. We learn to build better bridges only by assuming that collapses happen through natural causes that we can understand and learn from. In precisely the same way, it is the scientific quest for a natural understanding of life that has led to the discovery of relationships between species, genetics, DNA, and virtually all modern medicine. Many of the scientists who made these discoveries, including Charles Darwin himself, believed deeply that they could see God's hand in creation. But if they had let their belief stop them from seeking natural explanations, they would have discovered nothing. Intelligent design may or may not be true, and it may be worth discussing in philosophy classes. But if we allow science to be redefined to accommodate it, we will undermine everything that makes science so successful in advancing human knowledge.

Let's turn next to the second hallmark, where it is the criterion of simplicity that is most often misunderstood. To see why this idea is so important, you need only to remember that Copernicus's original model did *not* match the data noticeably better than Ptolemy's model. If scientists had judged Copernicus's model solely on the accuracy of its predictions, they might have rejected it immediately. However, many scientists found elements of the Copernican model appealing, such as the simplicity of its explanation for apparent retrograde motion. They therefore kept the model alive until Kepler found a way to make it work.

[2] When a similar discussion in a sidebar in my astronomy textbook drew complaints from a few ID proponents, I learned that some of them claim that the Designer need not be supernatural. But this is just semantics: If you believe that you've found evidence that life could not have evolved through the natural mechanisms of evolution, then by definition you are saying that a non-natural process intervened. To me, supernatural and non-natural are synonymous, but if you disagree, just substitute "a process that cannot be explained by Darwin's theory of evolution by natural selection." The meaning of my sentence won't change.

In fact, if agreement with data were the sole criterion for judgment, we could imagine a modern-day Ptolemy adding millions or billions of additional complexities to his Earth-centered model in an effort to improve its agreement with observations. In principle, a sufficiently complex model could reproduce the observations with almost perfect accuracy—but according to the way we view science today, the model still would not convince us that Earth is the center of the universe. We would still choose the Copernican view over the geocentric view because its predictions would be just as accurate yet would follow from a much simpler model of nature. The idea that we should prefer the simpler of two models that agree equally well with observations is often called *Occam's razor,* after the medieval scholar William of Occam (1285–1349). Like the idea that science should seek natural rather than supernatural causes, it is not any sort of absolute rule, but rather a guideline that has proven its value in the cause of scientific progress.

The third hallmark of science begs the question of what counts as an "observation" against which a prediction can be tested. To take us back to the main topic of this book, consider the claim that aliens are visiting Earth in UFOs. Proponents of this claim say that the many thousands of eyewitness observations of UFO encounters provide evidence that it is true. But should these personal testimonials count as *scientific* evidence? On the surface, the answer may not be obvious, because all scientific studies involve eyewitness accounts on some level. For example, only a handful of scientists have personally made detailed tests of Einstein's theory of relativity, and it is their personal reports of the results that have convinced other scientists of the theory's validity. However, there's a very important difference between personal testimony about a scientific test and an observation of a UFO: The first is at least in principle verifiable by anyone, while the second is not.

Understanding this difference is crucial to understanding what counts as science and what does not. Even though you may never have conducted a test of Einstein's theory of relativity yourself, there's nothing stopping you from doing so. It might require several years of study before you have the necessary background to conduct the test, but you could then confirm the results reported by other scientists. In other words, while you may currently be trusting the eyewitness testimony of scientists, you always have the option of verifying their testimony for yourself.

In contrast, there is no way for you to verify someone's eyewitness account of a UFO. Without hard evidence such as photographs or pieces of the UFO, there is nothing that you could evaluate for yourself, even in principle. (In the next chapter I'll discuss those cases where "hard evidence" for UFO sightings *has* been presented.) Moreover, scientific studies of eyewitness testimony show it to be notoriously unreliable. For example, different eyewitnesses often disagree on what they saw even immediately after an event has occurred; my own story at the beginning of this chapter is a case in point, since Grant and I have different versions of what happened to the flash of light as it disappeared from view. As time passes, memories of the event may change further. In some cases in which memory has been checked against reality, people have reported vivid memories of events that never happened at all. This explains something that virtually all of us have experienced: disagreements with a friend about who did what and when. Since both people cannot be right in such cases, at least one person must have a memory that differs from reality.

The demonstrated unreliability of eyewitness testimony explains why it is generally considered insufficient for a conviction in criminal court; at least some other evidence, such as motive, is required. And it is for the same reason that we cannot accept eyewitness testimony by itself as evidence in science, no matter who reports it or how many people offer similar testimony.

BEYOND UFOS

My personal UFO remains unidentified, leaving me free to believe what I want of it. If I want to, I can decide to follow my heart and imagine that I caught a glimpse of some of the intelligent beings who I really do believe share our universe with us. Or, I can keep my usual skepticism, and hold fast to my argument that it was more likely just a meteor.

And now I think you can understand the title of this book. No matter what I may believe about my UFO, there is nothing I can do to convince you that my belief is correct, especially if you are as skeptical as me. Some people think that makes skepticism bad, but I don't. It just means that instead of trying to convince you that aliens exist by telling you what I saw with my eyes, I need to go about it by concentrating on evidence that we can examine together. And that means we need to go beyond UFOs, and beyond arguments based solely on personal beliefs and opinions, and turn to science. Only through science will we actually learn something about other life in the universe, if indeed it exists.

3
WHAT I KNOW ABOUT ALIENS

Yet across the gulf of space, minds that are to our minds as ours are to those of the beasts that perish, intellects vast and cool and unsympathetic, regarded this earth with envious eyes, and slowly and surely drew their plans against us.

—*H. G. Wells, from* The War of the Worlds

From its title, you might expect this to be a very short chapter. I've never met an alien, I don't know that I'd recognize an alien if I saw one, and I'm not even sure that aliens exist. So what do I *know* about aliens? Nothing, absolutely nothing.

But you didn't really expect me to end the chapter quite so quickly, did you? I may not be able to say anything about aliens with 100 percent certitude, but as we discussed in chapter 2, science rarely can prove anything to be true beyond *all* doubt. Instead, science gives us a way to examine evidence and then to choose among possibilities. Sometimes the evidence for a particular possibility will become so overwhelming that we will regard it as truth. For example, that is the case for the idea of Earth going around the Sun rather than vice versa, an idea for which the evidence is so strong that it's difficult to imagine anyone seriously arguing with its reality.

We can also use science to consider various possibilities about aliens. If you knew nothing at all about life or the universe, you might guess that beings like us could live just about anywhere, including in space and on the Sun. Indeed, many ancient myths incorporated ideas much like this, since they often imagined mortals joining the gods among the stars. Today, based on what we know about the needs of life, we can be pretty confident that the best place to find life is on a planet or a large moon—an idea that we'll discuss in more detail in coming chapters. We can be similarly confident that we won't find beings traveling through space unless they first grew up on a planet or moon around some distant star, and then became smart enough to build spaceships. Those are the beings that I want to talk about in this chapter.

My knowledge of alien visitors to Earth is actually quite limited. I cannot tell you what they look like, or whether they have arms, legs, and eyes. I cannot tell you what their biochemistry is like, or whether their cells use DNA as genetic material. I don't know what they eat or breathe, or why they might be coming here, if indeed they are. But there's one thing that I *can* tell you: Technologically speaking, at least, they are very, very smart. If aliens really are visiting Earth and if they are drawing their plans against us—as in the excerpt from *The War of the Worlds* that opens this chapter—we don't stand a chance.

Now, I can cut a break here for H. G. Wells. The invaders of his novel came from Mars, a planetary neighbor that we ourselves can already reach with robotic spaceships. If there really were intelligent Martians, they wouldn't need to be much more technologically advanced than we are to plot their war against Earth. We'd presumably stand a fighting chance in such an invasion, especially if they had been lax enough in their study of biology to neglect the danger that earthly germs could cause them.

Hollywood doesn't deserve quite so much slack. The recent movie version of *The War of the Worlds* was vague about where the invaders came from, presumably since we now know that Mars is *not* home to an advanced civilization. Other movies with Earth invasions have been more direct in showing us fighting for survival against beings from the stars. Sorry, but we'd be squashed like bugs in any such battle, and that's not just my opinion. Rather, it's a conclusion that we can reach by scientifically examining the case for alien visitors, whom we may hope to be much less malicious than Hollywood usually thinks.

HOW SMART ARE THEY?
PART 1: EVIDENCE FROM SPACE

If you want to know how smart aliens would have to be to visit Earth, you need to understand how far they'd have to come. The best way to do that is to think about what space really looks like.

To our eyes, space looks crowded with stars, but this crowding is an illusion created by our lack of depth perception for very distant objects. Even stars that appear right next to each other in a constellation may in reality be separated by vast distances. I've devoted a large portion of my career to trying to explain just how vast those distances really are, and my method of choice is to start with our own solar system, and then move outward.

Suppose the Sun were the size of a grapefruit, so that you could hold it easily in your hands. How big would Earth be, and how far away would it orbit the Sun? Take a guess before I tell you the answer. Got your guess? Good. . . .

Most people guess that Earth would be something like the size of a golf ball on this scale, orbiting a few feet from the grapefruit-size Sun. These "most people" are right in realizing that Earth is smaller than the Sun, but pretty far off in realizing just how much smaller. In fact, with the Sun the size of a grapefruit, Earth would be no larger than the ballpoint in a pen, and would orbit far enough from the Sun for almost two first downs in football. The Moon, even smaller in size, would orbit only about an inch and half from the ballpoint-size Earth. In other words, the entire Earth–Moon system could fit with ease inside the Sun, and on this scale you can hold both Earth and the complete orbit of the Moon in the palm of your hand.

We can give the model a bit more precision by using an exact scale. The model sizes that I've described turn out to be almost precisely *one ten-billionth* of the actual sizes, so we'll use a scale of 1 to 10 billion. If you want to know any planet's size or distance from the Sun on this scale, you simply need to look up the real value and divide it by 10 billion. If you're interested in the numbers, this means the model Sun is 14 centimeters (5.5 inches) in diameter, Earth is 1.2 millimeters (1/20 inch) across, and Earth orbits at a distance of 15 meters (16.5 yards) from the Sun. Even more remarkably, if you calculate the circumference of Earth's orbit, you'll find that it is slightly greater than the length of a football field. So to envision our planet and its moon in space, just imagine holding the ballpoint-size Earth and even smaller Moon in your hand (careful not to drop them—they're important to us!), and taking a year to walk a distance as long as a football field as you carry them through their orbit around the grapefruit-size Sun.

What about the rest of the planets? Mercury is smaller than Earth and Venus is roughly the same size as Earth, so you can think of them as two more tiny ballpoints orbiting between the Sun and Earth. Mars is about half the size of Earth, and located another 8 yards from the Sun. The next four planets are somewhat larger—Jupiter and Saturn are the size of

Figure 3.1. This photo shows the pedestals housing the Sun (the gold sphere on the nearest pedestal) and the inner planets in the Voyage scale model solar system. The building at the left is the National Air and Space Museum. The model planets are encased in the sidewalk-facing disks that are visible at about eye level on the planet pedestals. (Photo by the author)

marbles on this scale, and Uranus and Neptune are peas—and are spread much farther apart. Jupiter orbits the Sun five times as far as away as Earth, Saturn is twice that far, and Uranus twice that far again. By the time you got to Neptune, you'd have walked more than a quarter mile from the Sun.

The best way to picture all this is to visit (or at least imagine visiting) an actual scale model of the solar system. There are quite a few such models around the world, but I know of only two that use the 1-to-10 billion scale. These are the two models that I've had a hand in building. The first, built with the help of my thesis advisor Tom Ayres and several undergraduate students, is on the campus of the University of Colorado, Boulder (starting outside the Fiske Planetarium). The second, pictured in figure 3.1 and built under the leadership of my friend Jeff Goldstein, is on the National Mall in Washington, D.C. This model, called *Voyage*, has its inner solar system just outside the east end of the National Air and Space Museum.[1] The outer planets are found along the walkway to the west, with Neptune and Pluto near the Smithsonian Castle.[2]

The first thing you'll notice as you walk through a scale model solar system is that the planets are really tiny compared to the spaces between them. To show the complete orbits of the planets around the Sun (rather than just showing the planets arranged in a straight line), the Voyage model would require an area equivalent to more than 300 football fields arranged in a grid. Spread over this large area, only the grapefruit-size Sun, the planets, and a few moons would be big enough to see with the naked eye. Perhaps you're beginning to see why we call it *space*.

Just to drive the point home, let me tell you a few more things you can learn as you imagine walking through a model solar system. First, think about the fact that while the solar system (out to Pluto) would require an area of 300 football fields in our model, the farthest a human being has ever traveled is the inch and a half to the Moon. Twelve people landed on the Moon as part of the Apollo program between July 1969 and December 1972. Next, to understand why a human trip to Mars would be so much more challenging, compare the distances of the Moon and Mars. At its closest, Mars is still about 150 times as far away as the Moon, and most of the time it is much farther. Using current technology for a trip timed to closest approach, your trip to Mars would take four to six months. Then, because Mars lines up with Earth in its orbit only about every 26 months (which is when it becomes brightest in our night sky), you'd have to stay there nearly two years until Mars again was close enough for your return trip home. For one more example, consider a trip to Pluto, the primary target of the *New Horizons* spacecraft. *New Horizons,* launched in January 2006, is traveling faster than any other spacecraft ever built. But even with

[1] *Voyage* was created as a joint project by NASA, the Smithsonian Institution, and the Challenger Center for Space Science Education. Our goal is to replicate the model at 100 or more other science museums and universities, so that people everywhere can learn to appreciate the scale of the solar system. If any readers are interested in sponsoring such a model (cost is approximately $180,000, including construction, installation, and educational materials), please contact me and I'll put you in touch with the right people.

[2] You've probably heard that the International Astronomical Union demoted Pluto to being a "dwarf planet" at their 2006 meeting. But Pluto is still part of the *Voyage* model, and we have no plans to bulldoze it—especially since the official definitions may yet change again.

its high speed, it won't reach Pluto until 2015. In other words, while you can walk the model distance from Earth to Pluto in just a few minutes, the real trip with current spacecraft technology will take nearly a decade. Finally, it's worth noting that the true scale of the solar system is so awesome that even professional astronomers are surprised. A typical reaction when I've taken professional astronomers on tours of the model solar system is "that can't be right," followed by a moment when they do the calculations in their heads and realize that it is. So you see, even for people who spend their lives studying this stuff, space is still much bigger than they usually think.

Indeed, when you think about the distances involved, it's quite amazing to realize that we've now sent spacecraft to photograph all of the planets in our solar system and many of their moons (as well as several asteroids and comets). But if aliens are visiting Earth, they don't come from around here. Our spacecraft photographs make it quite clear that no advanced civilization exists anywhere else in our own solar system. Visiting aliens must come from the stars, so the next question to ask is this: Using our scale of the model solar system, where are the nearest stars?

When I ask this question of children, typical answers start at about a mile away. A few might guess several miles. Then a child will shout out "100 miles!" and everyone will laugh, thinking it is so silly. But it's not. Imagine that you start at the Voyage model Sun in Washington, D.C. You walk the roughly 1/3-mile distance to Pluto and then decide to keep going to find Alpha Centauri, the nearest star besides the Sun.[3] The Voyage model doesn't actually include Alpha Centauri, but if it did, you'd want to bring a good backpack with plenty of food and several pairs of shoes, because you'd be walking to California. That's right: On the same scale where the Sun is a grapefruit with the ballpoint-size Earth orbiting just 15 meters away, you'd have to cross the United States to find the very next grapefruit. And that's just the model distance to the *nearest* star. Most stars are much, much farther away.

Given the vast distances to the stars, it may seem surprising that we can see them at all. After all, you wouldn't expect to be able to look from Washington, D.C., and see a grapefruit in San Francisco, even if you neglect the problems introduced by the curvature of the Earth. But we're talking about really bright grapefruits! Alpha Centauri is just as bright as

[3] If you are aware of and concerned about the fact that Alpha Centauri is actually a three-star system, just presume that I'm talking here about Alpha Centauri A, the brightest of the three stars and hence the one that we see with the naked eye.

the Sun, which is why we can see it as a tiny dot of light in the night sky despite its tremendous distance. This idea should also give you some perspective on seeing *planets* around other stars. If Alpha Centauri were orbited by a planet just like Earth, seeing this planet would be like trying to look across the United States to a ballpoint shining only with light that it reflects from the really bright grapefruit that it orbits. A planet like Jupiter would be slightly easier, but still like looking through a telescope for a marble some 2,500 miles away and hidden in the glare of a very bright street light. When I wish to be amazed about modern technology, I think about the fact that we can now detect such planets around nearby stars (though as we'll discuss in chapter 8, usually through their gravitational effects on their stars, rather than by actually "seeing" them), and that within a few years we should be capable of discovering planets as small as Earth.

The distances to the stars are so vast that ordinary distance units, such as miles or kilometers, cannot do them justice. Instead, we describe stellar distances in units of *light-years*. One light-year is defined to be the *distance* that light can travel in one year. If you think about how fast light moves, you'll realize that a light-year is a very long way. Light travels through space at a speed of 300,000 kilometers per second, or equivalently 186,000 miles per second. Stop and think for a moment about racing a light beam from your flashlight, and you'll realize that you just made a big mistake: During the second or so that you paused, the light could have circled eight times around the Earth. Light from the Sun reaches Earth in about eight minutes; the same trip in a car traveling 65 miles per hour would take you more than 160 years. The trip to Pluto, which will take the *New Horizons* spacecraft nearly a decade, is completed by light in under six hours. In a year, well . . . if you do the calculation by multiplying the speed of light by the number of seconds in a year, you'll find that a light-year is about 10 trillion kilometers, or 6 trillion miles. But it takes several years for light to reach even the nearest stars.

Alpha Centauri is a little over four light-years away, meaning that in the unlikely event that it exploded tomorrow, we wouldn't see the explosion until more than four years from now. Betelgeuse, the star at the upper left shoulder of Orion, is about 425 light-years away. Someone living on a planet around Betelgeuse with a telescope powerful enough to see the doings of people on planet Earth could at this moment be watching Tycho Brahe at work, collecting the data that would ultimately help us learn that Earth is not the center of the universe. On a planet orbiting Deneb, one of the three bright stars that make up the pattern known as the summer triangle, astronomers would see Earth at the time of the early Greeks, before the debate between the atomists and Aristotle had even begun.[4]

With all that in mind, we're ready to get back to our question of alien intelligence. The very minimum technology required for aliens to travel from their home stars to here might look something like a huge, multigenerational spaceship. Using nuclear rocket technology not too far beyond our own, they might be able to achieve speeds of a few percent of the speed of light, so that they could make the trip in only a few centuries if they are coming from nearby stars. But a ship like this would simply drift past Earth, so that if we saw it at all we'd see it as a steadily moving light for the few seconds during which it might be near enough to be seen. To explain the reports we hear of UFOs, the alien technology would have to be far more advanced.

Alien visitors of the type that people claim to see in UFO sightings are able to travel quite easily among the stars. They appear often, and in substantial numbers, and apparently don't mind visiting for just a few minutes or hours at a time even after the long journeys from the stars. It would seem that, for them, a journey to Earth is little more troubling than an intercontinental flight is for us, and certainly no more difficult than it is for us to reach the Moon. Let's put this in terms of our model solar system. Remember that, on our 1-to-10 billion scale, both Earth and the Moon fit in the palm of your hand, while the nearest stars are thousands of miles of way. To scale, they can cross a continent while we can barely travel an inch.

And there you have it: Alien visitors can in essence do the equivalent of flitting back and forth across the United States as easily as you can run your finger around in the palm of your hand. I have no idea what kind of technology they might have that would enable them to do that. What I *do* know is that this technology is far beyond what we have, and very likely beyond what we can even yet conceive of.

[4] I can't resist adding one more note: You know how people used to tell you to behave, because bad behavior would go on your "permanent record"? Well, get this: Suppose little Johnny just hit Suzie, but thinks he got away with it because you didn't catch him in the act. You might remind him that while you can't prove he did it, someone else might, because the image of his action is traveling outward through space at the speed of light. Technologically it might be a bit unrealistic, but in principle, 50 million years from now, beings living in a galaxy 50 million light-years away could point their sufficiently powerful telescope at Earth and catch Johnny in the act. So next time you hear someone say, "The truth is out there," you'll know that it really, truly is.

HOW SMART ARE THEY? PART 2: EVIDENCE FROM TIME

So far, I've shown that aliens would have to be very advanced to visit us, considering how far they would have to travel. But this distance argument is only part of the full equation. I'll now show you that when we consider time, we come to expect the aliens to be even more technologically advanced.

The time argument is a little more subtle, and it begins with our modern understanding of the age of the universe and the age of the Earth. I can just tell you the answers: The evidence shows that our universe is about 14 billion years old, while Earth is about 4 ½ billion years old. If you take my word on these ages, then I can complete an argument showing you that we should expect any visiting aliens to have technology that is at least tens of thousands of years beyond ours. But polls show that at least half of the public does not believe it when scientists state these ages, so it would not be any more fair for me to ask you to take my word for them than it would be for you to ask me to take your word about a UFO sighting. So if you'll bear with me for a bit, I will explain how the scientific evidence points to these old ages for Earth and the universe, and then return to the implications to alien technology.

Let's start with Earth. There are actually quite a few ways to put at least some rough limits on the minimum age of the Earth. For example, you can measure the rate at which a river causes erosion, and then estimate how long it must have taken to make the huge canyon that it runs through. These types of estimates will be very imprecise, but they show that our planet is quite old. Even some of the ancient Greeks guessed that Earth was at least hundreds of thousands to millions of years old, and by Darwin's time the geologists were quite confident that Earth's age was at least hundreds of millions of years. Today, we can measure Earth's age with remarkable precision through a technique known as *radiometric dating*.

To understand the technique, you need to know just a little about atoms. Atoms are made of protons, neutrons, and electrons, with the protons and neutrons bound together in what we call the *nucleus* of the atom. Every different chemical element has a different number of protons in its nucleus. For example, hydrogen has 1 proton, helium has 2 protons, and carbon has 6 protons. The number of neutrons can vary; in the case of carbon, the nucleus may contain either 6, 7, or 8 neutrons. The diagrams in figure 3.2 show the relevant terminology: The number of protons is called the *atomic number*, the combined number of protons and neutrons is called the *atomic*

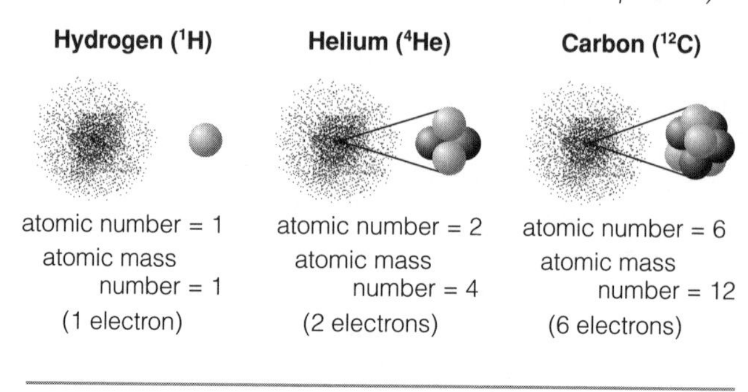

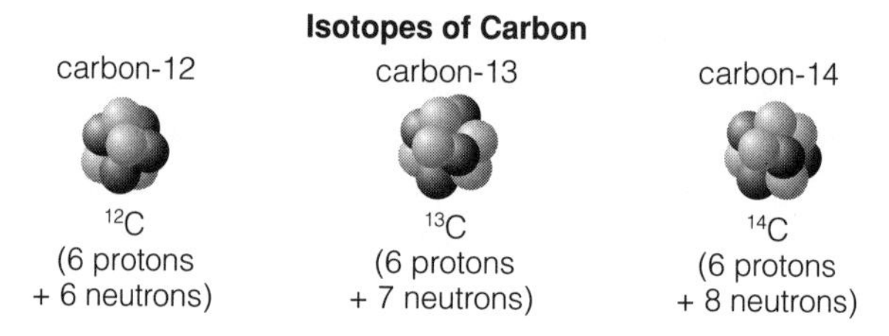

Figure 3.2. Terminology of atoms. (Illustration courtesy of Addison Wesley, an imprint of Pearson Education.)

mass (or, more technically, the *atomic mass number*), and atoms of the same element with different numbers of neutrons are called *isotopes* of one another.

The key to radiometric dating lies in the fact that some isotopes are "radioactive," which is just a fancy way of saying that their nuclei tend to undergo some type of spontaneous decay (such as breaking into two pieces or having a neutron turn into a proton), and that laboratory studies allow us to measure their precise rates of decay. Let's take a specific example that happens to be very useful in determining the age of the Earth. The radioactive isotope potassium-40 happens to turn into stable argon-40 when it undergoes decay. While the decay of any single nucleus is an instantaneous event, laboratory studies show that a large collection of potassium-40 atoms decay according to a very clear pattern. You don't have to watch for all that long (typically months or years) before you'll know the pattern, and you can use the decay pattern to calculate what we call the *half-life*: the time it would take half the atoms in any large collection to undergo decay. For potassium-40, the half-life turns out to be a fairly long 1.25 billion years.

Now, imagine a rock that has just solidified from molten lava and happens to contain, say, 1 microgram of potassium-40. Once solidified, the potassium-40 atoms are locked in place, which means that over time, you'd gradually find the potassium-40 atoms being replaced by argon-40 atoms. More precisely, because of its half-life, you'll find that after 1.25 billion years the 1 microgram of potassium-40 will have become ½ microgram of potassium-40 and ½ microgram of argon-40. After another 1.25 billion years, making 2.5 billion years total, you'll have ¼ microgram of potassium-40 and ¾ microgram of argon-40. And so on, with in-between times having in-between amounts.

I think you can now see the essence of radiometric dating. Suppose you find a rock that contains equal numbers of atoms of potassium-40 and argon-40. If you *assume* that all the argon came from potassium decay, then it must have taken precisely one half-life for the rock to end up with equal amounts of the two isotopes. You could therefore conclude that the rock is 1.25 billion years old. The only question is whether you are right in assuming that the rock lacked argon-40 when it formed. Answering this question about the original content of the decay product is usually the most difficult part of radiometric dating, but in this case it is pretty easy. Potassium-40 is a natural ingredient of many minerals in rocks, but argon-40 is a gas that never combines with other elements and that could not condense in the cloud that gave birth to our solar system. Therefore, if you find argon-40 gas trapped inside minerals, you can be confident that it came from radioactive decay of potassium-40.

Radiometric dating is possible with many other radioactive isotopes as well. For example, rocks that the *Apollo* astronauts brought back from the lunar highlands contain minerals with a very small amount of uranium-238, which decays (in several steps) into lead-206 with a half-life of about 4.5 billion years. Lead and uranium have very different chemical behaviors, and some minerals start with virtually no lead. Laboratory analysis of such minerals in lunar rocks shows that they now contain almost equal numbers of atoms of uranium-238 and lead-206. We conclude that half the original uranium-238 has decayed, turning into the same number of lead-206 atoms. The lunar rock therefore must be about one half-life old, or about 4.5 billion years old. More precise work shows these lunar rocks to be about 4.4 billion years old.

So how old is the Earth? The rocks we find on Earth's surface have a great variety of ages, since different rocks solidified at different times from molten lava. The very oldest Earth rocks are about 4 billion years old, but Earth itself must be older than this, because the entire surface has been reshaped through time. Mineral grains found within some Earth rocks date back as far as 4.4 billion years, about the same age as the Moon rocks, which must also be younger than Earth if, as currently thought, the Moon formed when a Mars-size object blasted part of Earth's outer layers into orbit (a hypothesis we'll discuss in chapter 5). The very best estimates of Earth's age come from radiometric dating of meteorites. The oldest meteorites are chunks of rock that must have solidified very early in the solar system's history, at the time when Earth and the other planets were just beginning to form. These meteorites are about 4.55 billion years old, which means that our planet began to form at about that time.

If you read a few Creationist web sites, you may find people trying to argue that radiometric dating doesn't really work. But as you've just seen, the scientific principles behind it are quite simple and the basic physics is very well understood: From our theory of nuclear structure we can determine which nuclei will decay and which will remain stable, and measured decay rates are in good accord with theoretical predictions. The technique of radiometric dating gains further support from the fact that, in many cases, rocks contain more than one type of radioactive isotope. If radiometric dating weren't valid, you might expect that different isotopes would give different ages for the same rock. But they don't: Within margins of measurement uncertainty, we invariably find that dating the same piece of rock with several different isotopes gives the same age, adding to our confidence in the validity of the technique. Moreover, we can often check the results we obtain from radiometric dating against other methods of measuring ages, giving us even more confidence that the technique works. For example, with many fairly recent archaeological artifacts, we can confirm the ages obtained from radiometric dating by measuring ages from tree ring data or even from dates written on the artifacts. In addition, astronomers have discovered completely independent methods for estimating the ages of stars, including the Sun. Although none of these methods give an age as precise as the age we obtain from radiometric dating, they all confirm that the Sun's age is in the range of 4 to 5 billion years. Overall, the technique of radiometric dating has been checked in so many ways and relies on such basic scientific principles that there is no longer any serious scientific debate about its validity.

With Earth's age under our belts, let's turn our attention to the age of the universe. Again, astronomers today have many independent ways of estimating this age. For example, the same stellar models that help confirm the age of the Earth enable us to estimate the ages of the oldest stars in the universe, which turn out to be about 13 billion years old. But the real key to estimating the universe's age lies with a discovery made in 1929 by Edwin Hubble (for whom the Hubble Space Telescope was named).

When Hubble began his work at the Mount Wilson Observatory in 1919, no one even knew whether other galaxies existed. Galaxies could be seen in telescopes, but astronomers were not sure whether they were gigantic and faraway collections of stars or small and relatively nearby clouds of interstellar gas. The ones that we now know to be spiral galaxies were dubbed "spiral nebulae," a name taken from their shapes and the Latin word for "cloud." Indeed, Pierre-Simon Laplace, whom we jointly credit with Immanuel Kant for the idea that our solar system was born from the collapse of a cloud of gas (see chapter 1), incorrectly guessed that the spiral nebulae were solar systems undergoing this birth process. Interestingly, Kant had a different idea. He thought that the birthplaces of stars lay in more diffuse nebulae, and we now know that he was generally correct in this belief. Meanwhile, he guessed that the dense-looking spiral nebulae were distant collections of stars that he dubbed "island universes"—and that we now call galaxies. Kant was very good guesser.

But guesswork isn't good enough in science, and until Hubble there was no conclusive evidence with which astronomers could settle the debate regarding the nature of the spiral nebulae. At Mount Wilson, Hubble had access to what was then the largest telescope in the world and the dark skies outside the then-small town of Los Angeles. By 1924, he had collected enough data to prove that the spiral nebulae consisted of individual stars and were indeed located far beyond the Milky Way. The debate was over, and we suddenly learned that rather than a universe that extended only as far as stars of the Milky Way, we live in just one of the many billions of galaxies that make up the cosmos. In one fell stroke, Hubble had discovered the universe to be some 100,000 times larger than anyone had previously imagined.

Over the next few years, Hubble not only continued to expand human knowledge, but he discovered that the universe itself is expanding. By measuring the distances and speeds of many galaxies, he uncovered what we now call *Hubble's law*: With the exception of a few nearby galaxies, all galaxies in the universe are moving away from us, and the farther away they are, the faster they're going.

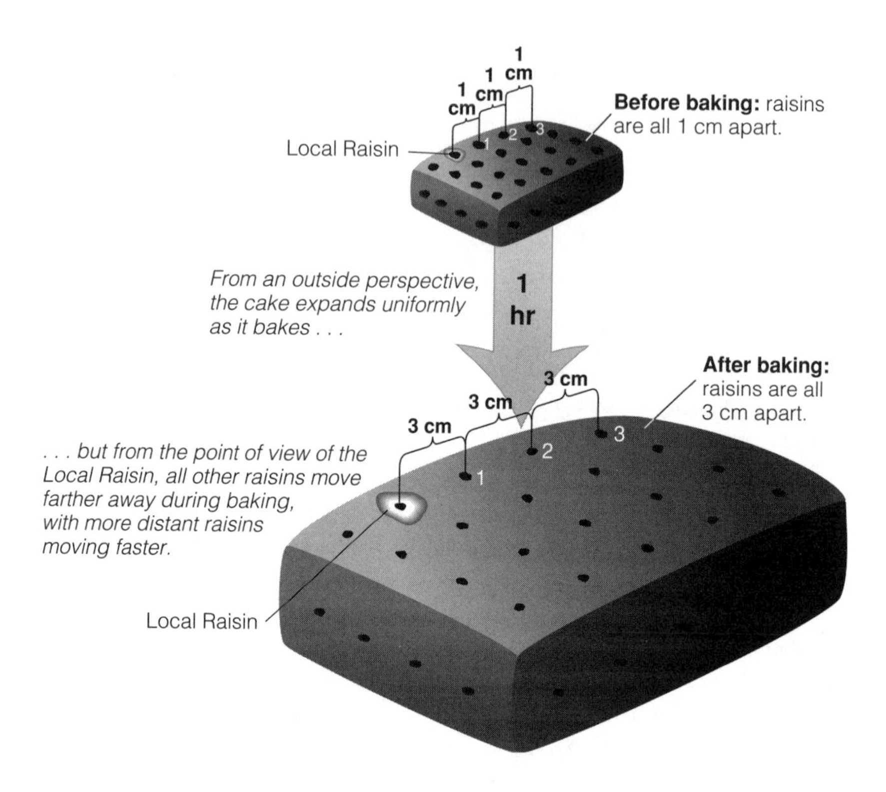

Distances and Speeds as Seen from the Local Raisin

Raisin number	*Distance before baking*	*Distance after baking (1 hour later)*	*Speed*
1	1 cm	3 cm	2 cm/hr
2	2 cm	6 cm	4 cm/hr
3	3 cm	9 cm	6 cm/hr
⋮	⋮	⋮	⋮

Figure 3.3. An expanding raisin cake offers an analogy to the expanding universe. (Illustration courtesy of Addison Wesley, an imprint of Pearson Education)

A simple analogy will show you how Hubble's law implies that the universe is expanding. Imagine a raisin cake in which you've carefully placed all the raisins exactly 1 centimeter apart. As shown in the diagram in figure 3.3, you place the cake into an oven, and over the next hour it expands until all the raisins are 3 centimeters apart. From the outside, the expansion of the cake is pretty obvious. But suppose you were a little raisin person living inside the raisin cake; how could you tell that your cake is expanding?

The diagram and table in figure 3.3 show the answer. Pick any raisin (it doesn't matter which one), call it your Local Raisin, and then notice what you would observe as you looked to other raisins before and after baking. Raisin 1, for example starts out at a distance of 1 centimeter before baking and ends up at a distance of 3 centimeters; this means that you'd see it move a distance of 2 centimeters during the hour of baking, which means it must appear to be moving away from you at a speed of 2 centimeters per hour. Raisin 2 starts out at 2 centimeters and ends up at 6 centimeters after the hour, which means it has moved a distance of 4 centimeters and hence has a speed of 4 centimeters per hour—twice as fast as Raisin 1. The pattern should already be clear: If you live inside an expanding raisin cake, you'll see all the raisins moving away from you, with more distant raisins moving faster. That is exactly what we observe galaxies to be doing and hence is how we know that the universe is expanding.

The fact of expansion almost immediately leads us to an estimate of the universe's age. The raisin cake in our analogy started out with the raisins already separated by 1 centimeter, but in the real universe we would not expect that type of arbitrary starting point. Instead, we assume that the current expansion implies that the universe was ever-smaller in the past, and that the expansion must have begun at some long ago moment in time when everything in the universe was all squeezed as closely together as possible. We call that moment the Big Bang. To see how we estimate when the Big Bang occurred, imagine that you thought the raisin cake had begun with its own Big Raisin Bang. Look again at the diagram and table, and notice that at the first time shown, Raisin 1 is 1 centimeter away and moving away from you at a speed of 2 centimeters per hour. Because it takes only ½ hour to travel 1 centimeter at this speed, you'd conclude that Raisin 1 must have been right on top of you ½ hour ago, which therefore must be when the Big Raisin Bang occurred. Note that you get the same answer no matter which raisin you choose. For example, Raisin 3 is initially 3 centimeters away and moving at 6 centimeters per hour, which again leads to the conclusion that it was right on top of you just ½ hour ago. In precisely the same way, Hubble's law tells us the rate at which galaxies are moving apart, so it is simple to work backward to estimate when the expansion started.

The only complication in doing this calculation is that we've been assuming that the rate of expansion now is the same as the rate that it was in the past. In reality, we don't expect that to be the case. We would expect that gravity would slow the expansion over time. However, in an example of why we should never go by intuition alone in science, astronomers have recently discovered strong evidence that the reality is precisely the opposite: Rather than slowing down, the expansion has actually been speeding up. The quest to understand this surprising discovery is now a full-time job for many of the world's leading scientists . . . but it is not the topic of this book.

Here, the important point is that we now can measure how the expansion rate has changed with time, and combining that information with the current expansion rate allows us to pin down the age of the universe with rather astonishing precision, given the task at hand. And the answer is: The universe is 13.7 billion years old, give or take a few hundred million years. We'll call it 14 billion years, which is close enough and well within the range of accuracy. Again, doubters may try to tell you that this is all a misinterpretation of the data; even a few dissidents within the astronomical community have tried to claim that the Big Bang didn't really happen. I'll admit that our confidence level in the Big Bang is not quite as high as it is for, say, the age of the Earth. But it's still quite high—personally, I'd put it at 99 percent—because the idea of the Big Bang is supported by several independent lines of evidence. For example, by assuming that the Big Bang really occurred, scientists can successfully explain both the microwave background radiation that fills space and the overall proportions of the chemical elements in the universe. No competing model can claim the same success. Similarly, the age measured from expansion gains further support since it agrees so well with ages of very old stars: Just as we would expect, the oldest stars seem to be a few hundred million years younger than the universe itself. So while I wouldn't be entirely surprised if, a century from now, someone picks up this book and laughs at what we thought about the Big Bang, I'd bet at least 100 to 1 that the evidence for the Big Bang by then will be even stronger.

Writing a book can be a strange thing. As I sit here, I feel like I'm really talking to you. But I don't actually know who you are, and I can't see your expression to see whether I've really convinced you that we live on a 4 ½-billion-year-old planet in a 14-billion-year-old universe. I hope so, but if not, at this point I'm going to have to ask you to play along with me so we can get back to the aliens.

The ages we have found tell us that the universe predates our solar system by nearly 10 billion years. Moreover, as we discussed in chapter 1, other evidence shows that the elements necessary for life are scattered throughout the universe. The abundance of these elements has risen through time as they are produced in stars, and there's some debate about how high the abundance needed to be before Earth-like planets could form. Nevertheless, it seems pretty clear that, at least in the Milky Way Galaxy, Earth-like worlds with life could have formed any time during the past 9 or 10 billion years, and possibly even earlier. Even if we then assume that, as on Earth, *intelligent* life could not arise until a planet was at least 4 to 5 billion years old, we are still left with an astonishing conclusion: In principle, intelligent life and civilizations could have begun to appear in our galaxy at least 5 billion years ago, which means some time before the birth of our own solar system.

This conclusion is even more astonishing if we consider its implications to alien technology. Recall that our galaxy has at least 100 billion stars. We do not know how many of these stars have Earth-like planets and civilizations, but we need to start somewhere, so let's guess. Let's suppose that the odds of a star having a suitable planet that eventually gets a civilization are about the same as the odds of winning the lottery, roughly 1 in 1 million. These may seem like long odds, but do the math: A billion is 1,000 times as large as a million, and 100 billion is 100 times larger still. In other words, even if only 1 in 1 million stars has a planet that ever gives rise to a civilization, we would still expect there to have been some 100,000 civilizations among the 100 billion stars in the Milky Way Galaxy.

Although this "calculation" of 100,000 civilizations is essentially a wild guess, it gives us a useful starting point. Stars come in all different ages, so there's no reason to think that any two civilizations would arise at the same time in the galaxy. Instead, the arrival of new civilizations in the galaxy would presumably be random, somewhat like the popping of kernels of popcorn. And here's the kicker: If we assume that the 100,000 civilizations have arisen randomly over the 5 billion years during which civilizations seem to have been possible, then the average time between the arrival of one civilization and the next is 5 billion years divided by 100,000—which is 50,000 years!

Think about this. If the assumption of 1 in 1 million stars getting a civilization is correct, then over the past 5 billion years our galaxy should have seen a new civilization rising up somewhere about every 50,000 years. Human civilization is only a few thousand years old, so statistically speaking we are almost undoubtedly the newest civilization on the galactic block. And how far ahead of us are the others? Assuming they have survived to the present, we'd expect the very next youngest civilization to have a 50,000-year head start on us, and the next to have a 100,000-year head start, and so on. The oldest civilizations would by now have already been around for many billions of years.

It literally takes my breath away. Under very simple and scientifically credible assumptions, if aliens are really visiting us, then at minimum we should expect them to come from a civilization that predates us by some 50,000 years, and possibly one that predates us by millions or billions of years. Changing the assumptions hardly changes the conclusion. If we really are being visited, then civilizations have arisen somewhere. If the number of civilizations is more than the 100,000 we guessed, then the average time between civilizations would be little shorter, so maybe the youngest are only 10,000 years instead of 50,000 years ahead of us. If the number of civilizations is smaller than 100,000, then the average time between them would be longer than the 50,000 years we found. No matter how you look at it, visitors to Earth must be thousands or tens of thousands of years beyond us technologically.

ALIEN TECHNOLOGY

So what does alien technology look like? If aliens are really here, then we automatically know that they can make easy journeys among the stars, proving that they are capable of things far beyond what we can now achieve. But the time argument makes them even more powerful. Let's take our earlier assumptions, and suppose we are being visited by the youngest civilization besides our own, the one that we might expect to be some 50,000 years ahead of us. Then figuring out what their technology might look like requires only that we imagine what our own technology will be like if we continue to develop it for the next 50,000 years.

Except it's unimaginable. The human rate of technological development has been rapidly accelerating. If you read science fiction from 50 years ago, you'll find that the writers rarely got anywhere close to envisioning what our present technology would look like. For example, I know of no one who envisioned the sudden rise of the Internet. Extrapolating into the future, I'll wager that none of us has any real idea of what technology will look like just 50 years from now, let alone in 500 years, 5,000 years, or 50,000 years. All I can say is that the technology will be incredible, and I'd expect it to look no more like our present-day technology than silicon chips look like the tools chimps use to dig out termites. To quote the science fiction writer Arthur C. Clarke, "Any sufficiently advanced technology is indistinguishable from magic."

BEYOND UFOS

And now we come full circle back to *The War of the Worlds*. A battle between us and visiting aliens would, at best, be like a group of cavemen trying to hold off the United States Army. The mere fact that we are still alive therefore proves that no one with intentions on our world has visited us lately, because if they had, the world would already be theirs. You can therefore breathe a sigh of relief, because we can be quite certain that any current visitors must be either benign or benevolent.

But could we really know that they're here? In light of what we've learned about their technology, most claims of "evidence" for alien visitation look downright silly. Again, remember that we are talking about beings who we should expect to possess technology at least about 50,000 years beyond ours. If they chose to make their presence known to us, does it really seem possible that they'd decide to do it by drawing patterns in wheat fields? (See figure 3.4.)

Claims of debris from alien crashes are hardly more plausible. Indeed, if you assume that debris from places like Roswell really is from alien

Figure 3.4. Crop circles. Does this really seem plausible as the chosen method of communication by extraterrestrials with technology 50,000 years beyond ours? (Photo by Joze Pojbic)

spaceships, the most remarkable thing about it is that alien spacecraft material doesn't look all that different from ours. But it should. After all, the composites used in modern military aircraft don't really look a whole lot like the wood used to build chariots in ancient times. Given the rate at which materials science is advancing, I'd be shocked if spacecraft a century from now bear much resemblance to anything we use today. Rather than being evidence of alien visitation, odd pieces of metal seem far more likely to be proof of human origin.

Indeed, I find the entire idea of crashed aliens to be very hard to swallow. Even with our current, primitive technology, we manage to build aircraft that hardly ever crash. That's why most of us are far more concerned with terrorists than with accidents when we board an airplane. Aircraft of the future will be even safer, thanks to technologies already in the pipeline that should drastically reduce accident rates. So if you ask me how often aircraft or spacecraft will crash in 50,000 years, here's my answer: essentially never. Even if true perfection cannot be achieved, the idea that multiple alien spacecraft could have crashed on farms and military installations in just the past few decades makes no sense at all when you consider how far they've come and how long they've had to perfect their technology.

Finally, the writing process demands that I address one more common claim about aliens, because this really did just happen to me: As I was working on these last couple of paragraphs on my own computer, Microsoft Office crashed, and I lost several sentences that I had to retype after restarting the program. And this brings me to claims that the government developed our current computer technology by "reverse engineering" alien computers recovered from crashed spacecraft. For the record, Bill Gates is one of my personal heroes for the tremendous charitable work he is doing with his money, but as my personal experience shows, his software is not always reliable. So if you want to believe that modern computing is the result of reverse engineering of alien technology, you'd have to believe that aliens could traverse interstellar distances with Microsoft software. Sorry, but I don't think they'd make it. Moreover, we all know that software is rapidly improving as people write more and better code. Given their 50,000-year head start, is it really conceivable that aliens would have been using code that looked like what we had in the 1970s, when the reverse engineering supposedly began?

The bottom line is that virtually any claim of "hard" evidence of alien visitation quickly collapses under its own weight of implausibility. But absence of evidence is not evidence of absence, and I'll be the first to admit that it's still possible that aliens are visiting us. In fact, based on the arguments

I've made about the number of civilizations, it seems difficult to believe that they're not—which presents a very interesting paradox that we'll discuss in the book's final chapter. Meanwhile, we are left with a simple reality: With extremely high confidence, we can conclude that any aliens who are visiting Earth are so far beyond us that there's virtually no chance of them leaving evidence behind by accident. If they want us to know they're here, they'll tell us. And if they don't want us to know, it's highly unlikely that we could discover them, no matter how hard we might try.

And yet, I still won't tell anyone who claims to have seen a UFO that they're wrong. Because with all this considered, there are still laws of nature that must be obeyed. I suspect that aliens will have discovered laws of which we are unaware, perhaps allowing them to do things like "cloaking" their spacecraft to prevent us from seeing them. But if that kind of hiding isn't possible, and if they really are visiting, then no matter how good their technology, we might catch an occasional glimpse of one of their spacecraft as it darts by. I highly doubt that such craft would look at all like flying saucers or anything else that is commonly reported. But if you see a strange light in the sky, indistinct but unmistakable and moving in a way that defies ordinary explanation, then maybe, just maybe, you've seen them. Maybe even I've seen them, back when I saw my own UFO during the meteor shower.

So now, we really are ready to move beyond UFOs. I hope I've convinced you that even if UFOs are real, we probably won't be able to find the hard evidence that science demands to prove it. It's time to turn our attention to things that we *can* study through science, such as life on Earth, the possibility of life elsewhere in our solar system, and possibilities for life among the stars.

4

WHAT IS LIFE?

There is grandeur in this view of life, with its several powers, having been originally breathed by the Creator into a few forms or into one; and that, whilst this planet has gone cycling on according to the fixed law of gravity, from so simple a beginning endless forms most beautiful and most wonderful have been, and are being, evolved.

—*Charles Darwin,* The Origin of Species, *1859*

If intelligent aliens someday land in Times Square or Red Square or Tiananmen Square and announce themselves to the world, the implications to life in the universe will be immediately clear. Their advanced technology will prove they are from a distant world, and even if they themselves are robots or androids or genetically engineered organisms bearing little resemblance to their forebears, they'll still represent indisputable proof that life has arisen elsewhere. A SETI success in receiving a message from another civilization would be only slightly less dramatic, and equally profound in demonstrating that we are not alone in our universe. But unless or until something this spectacular occurs, our only realistic hope of finding life beyond Earth lies with using spacecraft or telescopes to search for it. And in that case, it's important to know exactly what it is that we are searching for.

You might think that it would be easy to define life, but it's not. Consider a cat and a car, which turn out to have a lot in common. Both require energy to function—the cat gets energy from food, and the car gets energy from gasoline. Both can move at varying speeds and can turn corners. Both expel waste products. But a cat clearly is alive, while a car clearly is not. What's the difference?

In the case of a cat and a car, we can find many important differences without looking too far. For example, cats reproduce themselves, while cars must be built in factories. But as we look deeper into the nature of life, it becomes increasingly difficult to decide what characteristics separate living organisms from rocks and other nonliving materials. Indeed, the question can be so difficult to answer that we may be tempted to fall back on the famous words of U.S. Supreme Court Justice Potter Stewart, who, in avoiding the difficulty of defining pornography, wrote: "I shall not today attempt further to define [it]. . . . But I know it when I see it." If living organisms on other worlds turn out to be much like those on Earth, it may prove true that we'll know them when we see them. But if the organisms are fairly different from those on Earth, we'll need clearer guidelines to decide whether or not they are truly "living."

THE DIVERSITY OF LIFE ON EARTH

Part of the difficulty in defining life arises from the fact that, even here on Earth, most life is quite different from the life we see all around us. If you went strictly by appearances, you'd probably conclude that all life belongs to one of just two kingdoms of life—the plant kingdom and the animal kingdom. But this "obvious" view of life, held in various forms for thousands of years, is just as misleading as the equally old and "obvious" idea that the Sun, Moon, and stars circle around us each day. In truth, we are no more the center of Earth's biological universe than Earth is the center of the physical universe.

The idea that our biological status is less impressive than we generally presume is much less well known than the idea that we live on one small planet in a vast universe, in part because the true breadth of biological diversity on Earth has only recently been recognized. Indeed, up until just a couple decades ago, most biology textbooks continued to emphasize the idea that the plant and animal kingdoms represented a large part of life on Earth. But as we now know, these kingdoms are "large" only in the sense that many of their members are big enough to see with the naked eye. In any other sense—such as in the total biomass they represent or in the importance of their roles in shaping Earth's atmosphere—they are much less dominant than other forms of life on Earth.

The first evidence that biological reality is different than it appears to the naked eye came around the same time that the Copernican revolution was concluding. Not long after Galileo turned his telescopes to the heavens, other scientists began to employ similar lens technology to study the world of the very small. The precise origin of the microscope is not known, but the first practical microscopes used for scientific study were built by the Dutch scientist Anton van Leeuwenhoek (1632–1723; last name pronounced lay-ven-hook). During decades of observations beginning around 1674, Leeuwenhoek discovered the world of microscopic life. He was the first to realize that drops of pond water are teeming with microorganisms—a discovery now repeated by almost every elementary school student.

The discovery of the microbial world came as a big surprise, but for a long time it did not fundamentally alter human perceptions of the biological universe. Microbes were thought of as tiny plants or animals; for example, bacteria were considered plants, while mobile protozoans, such as amoebas, were classified as animals. It was only in the 1960s that biologists generally accepted that these organisms really didn't fit the traditional two-kingdom classification scheme. The kingdom list therefore expanded from two to five, with two of the new kingdoms (monera and protista) reserved for microorganisms; the third new addition was fungi, such as mushrooms, by then recognized to be different from both plants and animals.

You can still find this five-kingdom scheme in many old or not-quite-up-to-date textbooks, but biological science has come a long way since then. As scientists began to learn more about the biochemistry of cells, they realized that microbes that look similar under a microscope often have striking biochemical differences, such as fundamental differences in the chemical structures of their cell membranes or of the proteins that regulate metabolism. More recently, the same type of technology that made possible the Human Genome Project—the project that determined the sequence of all 3 billion or so base pairs in human DNA (short for *d*eoxyribo*n*ucleic *a*cid)—has been used to determine the genetic relationships between many different species, including microbial species.

To understand the basic concept, it's helpful to review the way that DNA encodes hereditary information, by which we mean the "operating instructions" that a living organism passes on to its descendants. The molecular structure of DNA, a *double helix,* is one of the most familiar scientific icons of our time (see figure 4.1). A helix is a three-dimensional spiral, such as you would make by extending a Slinky toy; a double helix has two intertwined strands, each in the shape of a helix. The structure looks much like a zipper twisted into a spiral. The fabric edges of the zipper represent the "backbone" of the DNA molecule, while the zipper teeth that link the two strands represent molecular components called DNA bases. The chemical structure of the backbone is interesting and important in its own right, but it is the DNA bases that hold the key to heredity. All known life on

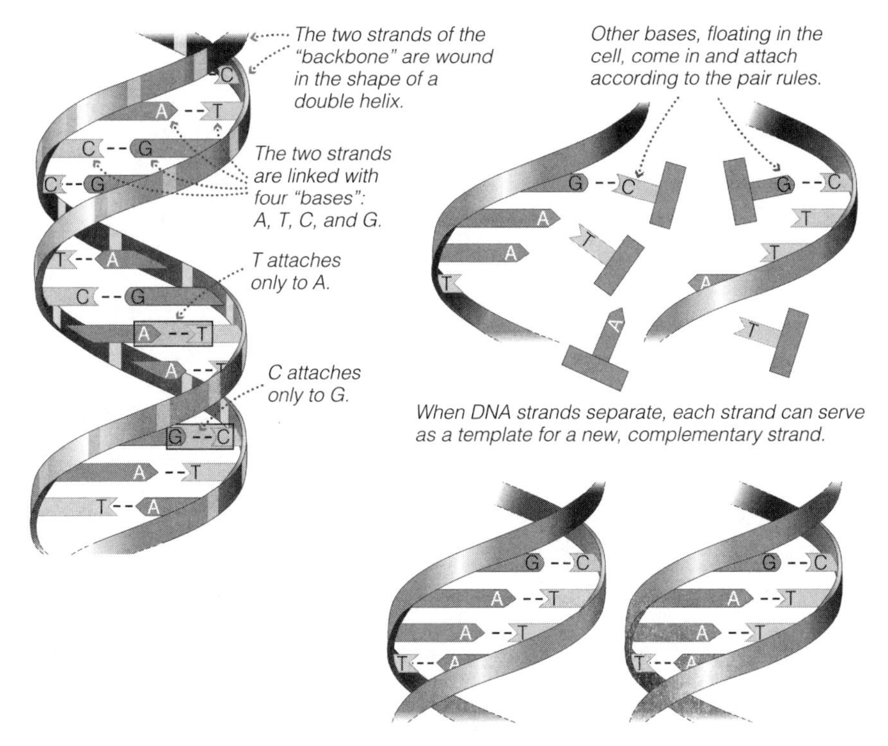

Figure 4.1. (Left) The double helix structure of a DNA molecule, which looks much like a zipper twisted into a spiral. The important hereditary information is contained in the "teeth," or DNA bases, linking the strands. Only four DNA bases are used, and they can link up only in specific ways: T attaches only to A, and C attaches only to G. (Right) These base pairing rules make it possible for a DNA molecule to be copied. (Illustration courtesy of Addison Wesley, an imprint of Pearson Education)

Earth makes use of only four DNA bases: adenine (abbreviated A), guanine (G), thymine (T), and cytosine (C).

In order to pass heredity information down through generations, living cells must be able to pass their DNA to their offspring. This is possible because of "base pairing rules" that dictate the way the two strands of DNA can be linked together: T can pair up only with A, while C can pair up only with G. Figure 4.1 shows the idea by representing the different bases with different shapes. For example, the shape of A, which is depicted as ending with an open triangle, fits only into the notch in T. Similarly, what is shown as the curved end of G fits only into the curved notch in C. These diagrams are only schematic representations—there aren't literally notches and curves of these shapes—but the real chemical bases work much the same way: Their actual chemical structures determine how they pair up.

The base pairing rules mean that when the two strands separate, new strands can be assembled by following the rules along each one. The process of DNA replication begins when other molecules in a living cell come in and "unzip" the double helix. Once it is unzipped, DNA bases floating freely within the cell can be brought in to pair up with each of the two separated DNA strands. Because these new bases must link to the bases on the existing strands according to the base pairing rules—T with A and C with G—they ultimately form complementary new strands that are joined to the existing ones. (By saying that two strands are *complementary,* we mean that one contains the precise sequence of bases needed to match the other in accord with the base pairing rules.) As long as no errors occur, the end result is two identical copies of the original DNA molecule. When a cell divides, one copy goes to each daughter cell, ensuring that the daughters have the same genetic information as the parent cell.

Besides having the ability to be replicated, DNA also determines the structure and function of the cells within any living organism. In essence, the "operating instructions" for a living organism are contained in the precise arrangement of chemical bases (A, T, C, and G) in the organism's DNA. Within a large DNA molecule, isolated sequences of DNA bases represent the instructions for a variety of cell functions. For example, a particular sequence of bases may contain the instructions for building a protein or for carrying out or regulating one of these building processes. The instructions representing any individual function—such as the instructions for building a single protein—make up what we call a *gene.*

Interestingly, among plants and animals, most of the DNA is not part of any gene; that is, it does not appear to carry the instructions for any particular cell function. For example, this so-called *noncoding DNA* (sometimes called "junk DNA") makes up more than 95 percent of the total DNA in human beings. Biologists suspect that most of this noncoding DNA represents evolutionary artifacts—pieces of DNA that may once have had functions in our ancestors but that no longer are important, much like the appendix is an organ that no longer plays an important role in our bodies. However, recent discoveries suggest that at least some of the noncoding DNA may play important roles that are not yet fully understood.

The complete sequence of DNA bases in an organism, encompassing all of the organism's genes as well as all of its noncoding DNA, is called the organism's *genome*. Different organisms have genomes that vary significantly both in total length (number of bases) and in their numbers of genes. For example, some simple microbes have DNA that extends only a few hundred thousand bases and contains only a few hundred genes.[1] We humans have a genome that contains an estimated 20,000 to 25,000 genes among its sequence of some 3 billion DNA bases. Note that, genetically speaking, we are by no means the most complex organisms on Earth. Rice, for example, has significantly more genes than we do—recent sequencing of the rice genome shows that it has about 37,000 genes, though it has a shorter total DNA sequence. Other organisms have far more DNA than people. For example, the simple plant known as the "whisk fern" (*Psilotum nudum*) has more than 70 times as many bases in its genome as humans, though most of this extra DNA is probably noncoding.

I've taken you on this little digression through molecular biology because you can now understand how biologists map genetic relationships between living species. Today, biologists have technology that allows them to determine the sequence of bases in almost any strand of DNA. This technology has been used to determine the DNA sequences that code for many cell functions, as well as to determine the complete DNA sequences of many living organisms, including humans. While there is always some variation among individuals, every member of a particular species has the same basic genome. By comparing the DNA sequences in similar genes among different species, biologists can determine how closely the species are related. For example, two species with very similar DNA sequences must be closely related, while two species with very different DNA sequences must be much more distant relatives.

And here's where the big surprise comes: Plants, animals, and fungi, which seem so different to us, are in reality far more closely related to one another than almost any two microbial species that you might pick up at random. Biologists show the genetic relationships between different types of organisms with the "tree of life," pictured in figure 4.2. The first thing you'll notice is that it's not your grandfather's tree, with just two large

[1] Many viruses are far simpler, with just a few thousand bases and a handful of genes. Mitochondria within plant and animal cells, which are thought to have had free-living ancestors, are also much simpler than the simplest bacteria sequenced to date. For example, human mitochondria have fewer than 17,000 DNA base pairs representing fewer than 40 genes.

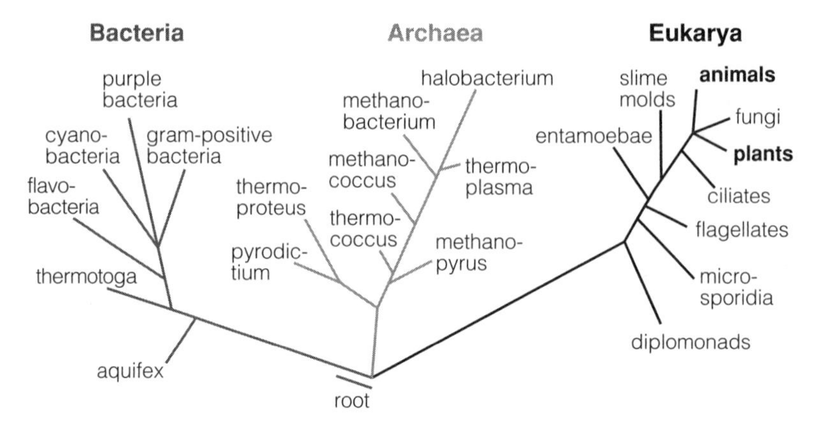

Figure 4.2. The tree of life, as it is known today, has three major domains: Bacteria, Archaea, and Eukarya. Note that all plants and animals represent just two small branches of the domain Eukarya. Only a few of the many known branchings within each domain are shown. (Illustration courtesy of Addison Wesley, an imprint of Pearson Education, developed with the assistance of the Pace Lab at the University of Colorado)

trunks for plants and animals. Instead, there are now three trunks, identified as *Bacteria, Archaea,* and *Eukarya,* and the entire plant and animal kingdoms are just two small, closely linked branches within the trunk labeled Eukarya. The three trunks are technically referred to as *domains,* though it may be easier to think of them as "superkingdoms." To summarize, we now know that there are dozens of different "kingdoms" of life, each with at least as much variety as we find within the plant and animal kingdoms, and that these kingdoms can be grouped into the three domains (or superkingdoms) Bacteria, Archaea, and Eukarya.

This new understanding of the tree of life has astonishing implications. It immediately shows that creatures like us are *not* the dominant forms of life on Earth, and that plants and animals come nowhere close to representing the full range of life forms even here on our own planet. Instead, the true diversity of life on Earth is found almost entirely within the microscopic realm. Moreover, because the vast majority of microscopic species probably remain undiscovered, we are likely to find much more diversity as we continue to study biology. We will certainly discover many more branches within the three domains, and it is even possible that entirely new domains will be discovered in the future.

This tremendous diversity among microbes extends beyond their genetics and into the places in which they live, a fact with important implications to the search for extraterrestrial life. In particular, we now know that the search for life need not be limited to a search for places with environments in which *we* could survive. We humans, along with all other animals and plants, need oxygen and a fairly narrow range of temperatures, water salinities, and other things to survive. But there are microbes living under a much broader range of conditions, including conditions likely to exist on many other worlds that we would hardly categorize as Earth-like. The diversity of these so-called *extremophiles* ("lovers of the extreme")—meaning microbes adapted to environments that are extreme by human standards—is truly remarkable, so let's consider just a few examples, starting with the *hyperthermophiles* that survive in conditions of unusually high temperature.

You probably know that the best way to kill bacteria when you are cooking is to make sure you raise the temperature high enough: You'll kill *salmonella* by cooking your meat to above about 180°F, and you'll make sure your soup is safe by bringing it to a boil at 212°F. These heat techniques work because, in general, DNA and other biologically important molecules fall apart at such high temperatures. And yet, in recent decades, scientists have discovered life—mostly microbes of the domain Archaea—thriving in the very hot water around *black smokers,* which are volcanic vents on the ocean floor (color plate 2, top). For example, an organism called *Pyrolobus fumarii,* whose name means "fire lobe of the chimney," was actually discovered *in* the walls of a black smoker, where it grows in water heated to as high as 235°F. (The water doesn't boil because it is kept liquid by the high pressure at the bottom of the ocean.) And in 2003, researchers discovered another species of Archaea living near the black smokers that can grow in even hotter water. This species does not yet have an official name but is being called "Strain 121" because it can grow in water as hot as 121 degrees Celsius, or 250°F; in addition, it can survive in the lab for up to two hours at temperatures of 266°F. Similar organisms thrive in hot springs on the Earth's surface, such as in the springs found in Yellowstone National Park (color plate 2, bottom).

Other extremophiles live in conditions far too cold, acidic, alkaline, or salty for "ordinary" life to survive, and some seem like they should be able to survive even on other worlds of our own solar system. For example, microbes called *endoliths* (meaning "within rocks") can live as much as a couple of miles below Earth's surface, as long as there are microscopic amounts of liquid water filtering through the subterranean rock. These microbial species live almost completely independent of the surface environment. They do not eat, instead getting nutrients from chemicals and air trapped within the rock, and they do not depend on sunlight in any way, since they get energy from chemical reactions between the liquid water and minerals in the rock. One community of endoliths, discovered in 1995, consists of bacteria living deep beneath the surface of Oregon and Washington in a rock formation known as the Columbia River Basalt. Other rock-dwelling microbes survive in the cold, dry valleys of Antarctica, and some are known to survive temperatures as low as about −4°F, as long as even a very thin film of liquid water is available. As strange as these microbes may seem, they are very common on Earth; indeed, some biologists suspect that the total mass of subsurface organisms living in rock may exceed that of all the life on Earth's surface. Moreover, as we'll discuss in chapter 7, the type of subsurface environment in which they live almost certainly exists in similar form on other worlds, including Mars.

The range of conditions in which various species can survive is truly astonishing, and new record holders for temperature, salinity, depth in rock, or other conditions are discovered almost every year. Still, one known species will be hard to beat for its sheer audacity: the bacterial species known as *Deinococcus radiodurans,* which can survive radiation more than 1,000 times that which would be lethal to humans and other animals. These remarkable organisms actually thrive in radioactive waste dumps! How do they do it, when the radiation can easily destroy the DNA upon which they depend for heredity? Amazingly, they apparently have cellular mechanisms that can repair the DNA faster than it is destroyed, an ability clearly of great interest to medical researchers. From the standpoint of astrobiology, *D. radiodurans* (as the species is known for short) is important because it could survive the radiation exposure on many worlds with atmospheres less protective than Earth's. Perhaps more significantly, if an asteroid hit our planet and blasted a rock into space with hitchhiking *D. radiodurans* aboard, the microbes might well survive long enough for their meteorite to land on Mars.[2] As we'll discuss in the next chapter, this idea raises the possibility that life might have migrated among the inner planets of our solar system in the past.

[2] Other microbes can survive trips through space by going into a "dormant" state. For example, the deadly bacteria anthrax can form "resting cells," or *endospores,* that can survive a complete lack of water, extreme heat or cold, and most poisons. Some endospores can remain dormant for centuries.

It's worth keeping in mind that while extremophiles may seem "extreme" to us, they would probably say the opposite if they could talk. For example, many hyperthermophiles die when brought to "normal" temperatures, because their enzymes have evolved to function only at the high temperatures in which they live. Indeed, many extremophiles are anaerobic (meaning they live without oxygen), and they are poisoned by the oxygen on which our own lives depend. And this leads me to another very important point about human capabilities.

I count myself as an environmentalist, and I'm very concerned about what we are doing to our planet and to many of the beautiful species that share it with us. But when I hear people claim that we might somehow wipe out life on our planet, I know it's pure hubris. Sorry, folks, but we just aren't that powerful. Sure, we could lay waste to our own civilization with nuclear bombs, and we might even change the climate enough to lead to our own demise as a species, sadly taking millions of other animal and plant species along with us. But the endoliths living miles beneath our cities in Oregon and Washington, the hyperthermophiles living near the black smokers, and *D. radiodurans* and others won't even bat a proverbial microscopic eye. We could destroy the ozone layer, raise the temperature, change the oxygen fraction, and radiate the surface with nuclear waste, and most life on Earth won't even notice.

The fact that we probably cannot harm most microbes has a positive side, because while microbes could care less about us, we depend heavily on them. Although microbes are most likely to make the news when they do something bad, like cause a disease, most microbes are harmless and many are crucial to our survival. For example, bacteria in our intestines provide us with important vitamins, and bacteria living in our mouths prevent harmful fungi from growing there. Other microbes play crucial roles in cycling carbon and other vital chemical elements between organic matter and the soil and atmosphere; for example, microbes are responsible for decomposing dead plants and animals. We even owe the existence of the oxygen we breathe to microbes; according to present understanding, Earth's atmosphere began essentially oxygen-free, and gained oxygen only as it was produced by tiny, photosynthetic bacteria. Without microbes, all plant and animal life would be doomed. In contrast, microbes could survive just fine without plants and animals, as they did during most of the history of life on Earth. We are not the center of Earth's biological universe, let alone of the rest of the universe.

WHAT LIVING THINGS DO

The great diversity of life on Earth might give you some pause when you think about the question of how to define and recognize life. After all, how would you know whether the rock you pick up contains microscopic organisms? For Earth-based biologists, however, this question isn't too difficult, because there are chemical tests we can do to look for life on Earth. For example, all known life on Earth uses DNA as its hereditary material, so we should find traces of DNA if a rock contains living organisms.

The question really only becomes difficult when we think about possibilities beyond Earth. Although life anywhere would presumably have some sort of hereditary material, we have no reason to expect that it would be DNA. (It's even possible that non-DNA life remains undiscovered right here on Earth, though it seems unlikely.) So if we picked up a rock from Mars or from the cold surface of Titan, how would we know whether it contained something that qualifies as life?

One way to answer this question is by making a list of the properties that seem to be common to life but independent of a particular biochemistry. In this way, we can look for characteristics that we might expect to be shared among all life forms, regardless of where they exist or how they arose. Different scientists have made such lists in many different ways, and they don't always fully agree with one another. Nevertheless, I believe you'll find general agreement that all life must share at least the following three properties:

1. Life creates order out of chaos.
2. Life makes copies of itself, but not quite perfectly.
3. Life evolves through time.

Let's consider each of these properties in a little more detail. The first expresses the fact that living organisms can take fairly random assortments of molecules and arrange them into orderly patterns that make cell structures and govern the metabolism of life. We sometimes find order in the physical world as well—for example, rock crystals are highly ordered—but the order comes about in a different way. Living organisms use energy to create the order upon which their survival depends. This is an important and often misunderstood concept that is related to something called the *second law of thermodynamics*. Thermodynamics is a branch of science that deals with energy and the rules by which it operates. The first law of thermodynamics, also known as the law of conservation of energy, tells us that energy can neither be created nor destroyed but only transformed from one form of energy to another. The second law of thermodynamics states that, when left alone, the energy in a system undergoes conversions that lead to increasing disorder (measured quantitatively as *entropy*). Living organisms are a perfect example of this law's importance: If you place a living organism in a sealed box, it will eventually use up the available energy and therefore no longer be able to build new molecules or fuel any of the molecular processes needed for life. Its molecules will therefore become more disordered with time—for example, the molecules may decay or may lose the orderly relationships they maintain with other molecules when the organism is alive—causing the organism to die. Thus, to maintain order and survive, a living organism must have a continual source of energy that it can use to counter the tendency for disorder to take over. As we'll discuss more later, this need for energy to maintain order helps constrain the list of places that seem possible abodes of extraterrestrial life.

The second property is what we generally think of as reproduction: Living organisms reproduce their own kind. Reproduction seems necessary to a definition of life because, without it, there would be no way for life as a whole to survive the deaths of individuals. However, if the reproduction were perfect, like the copying of digital computer files (hopefully!), all individuals of a species would always be exactly alike. This type of uniformity would be disastrous to life for a variety of reasons. For example, if all wheat were genetically identical, it would take just one infectious disease to wipe out all the wheat in the world, since there'd be no hope of finding a wheat plant with a genetic variation that could enable it to resist the disease. (Indeed, for just this reason agricultural scientists are concerned about the trend toward mass production of genetically similar crop strains.) Living organisms must reproduce themselves, but they must do it in a way that is not quite perfect so that there will always be individual variation. For life on Earth, this necessary imperfection comes from slight changes, or *mutations*, that sometimes occur in DNA, which is why the individuals within species are not all perfect clones of one another.[3]

The third property, that life evolves through time, is what makes it possible for species to adapt to changes in the environment around them. Without the ability for evolutionary adaptation, life probably would not be possible over the long haul. The physical environment inevitably undergoes change. Earth's climate has seen substantial changes over time, from

[3] Of course, we humans differ from our parents primarily because of the way genes are shuffled in sexual reproduction, rather than due to new mutations. But without mutations, sexual reproduction never would have evolved in the first place.

deep ice ages to very warm periods. Adaptation to such change has allowed life to survive and thrive on our planet. More important, we presume that life must have somehow got started long ago, with some ancestral microbes that were the first to be able to create order from chaos and reproduce on our planet. Without evolutionary adaptation, these microbes would never have changed and blossomed into the diversity of life that we find today.

If you think about it, these three properties form a simple hierarchy. The first property, creating order out of chaos, is implicit in the second, the ability to reproduce. And the second, not-quite-perfect reproduction, is implicit in the fact that life is able to evolve through time. And that is why, despite the public controversy over evolution, biologists today recognize evolution as the central, unifying theme of all biological science. Indeed, if you're looking for a simple definition of life, this will probably suffice: *Life is a chemical system with the ability to evolve through time.*

UNDERSTANDING EVOLUTION

Evolution is probably the single most misunderstood idea in all of science, which probably explains why so many people do battle over it. After all, it's a lot easier to argue with something by drawing caricatures of it than by actually digging in to understand it. But like most of the great, unifying principles of science, the theory of evolution has an underlying simplicity that anyone can appreciate with just a little bit of effort. Because evolution is so important to defining life, and hence to the search for life in the universe, I hope you won't mind if I spend a few pages in an attempt to demystify it.

The word *evolution* simply means "change with time," and the idea that life might evolve through time goes back more than 2,500 years. The Greek scientist Anaximander (c. 610–547 B.C.) promoted the idea that life originally arose in water and gradually evolved from simpler to more complex forms. A century later, Empedocles (c. 492–432 B.C.) suggested that creatures poorly adapted to their environments would perish, foreshadowing the modern idea of evolutionary adaptation. Many of the early Greek atomists probably held similar beliefs, though the evidence is sparse. Aristotle, however, maintained that species are fixed and independent of one another and do not evolve. This Aristotelian view eventually became entrenched within the theology of Christianity, with the result that evolution was not taken seriously again for some 2,000 years.

By the mid-1700s, scientists were beginning to recognize that many fossils represented extinct ancestors of living species, and the idea that Earth was quite old was gaining widespread acceptance. However, no one yet knew *how* species might change with time. For that, we needed a model that could describe the mechanism of evolutionary change.

The first serious model for evolution was proposed in the early 1800s by the French naturalist Jean Baptiste Lamarck. He proposed a mechanism known as "inheritance of acquired characteristics," in which he suggested that organisms develop new characteristics during their lives and then pass these characteristics on to their offspring. For example, Lamarck would have imagined that weightlifting would enable a person to create an adaptation of great strength that could be genetically passed to his or her children. While this hypothesis may have seemed quite reasonable at the time, it has not stood up to scientific scrutiny and, following the scientific practices I outlined in chapter 2, it has therefore has been discarded as a model of how evolution occurs. It has been replaced by a different model, proposed by the British naturalist Charles Darwin.

Charles Darwin described his theory of evolution in his book *The Origin of Species,* first published in 1859. In this book, Darwin laid out the case for evolution in two fundamental ways. First, he described his observations of living organisms, made during his five-year voyage on the HMS *Beagle,* and showed how they supported the idea that evolutionary change really does occur. Second, he put forth a new model of how evolution occurs, backing up his model with a wealth of evidence. In essence, the fossil record and the observed relationships between species together provide strong evidence that evolution *has* occurred, while Darwin's theory of evolution explains *how* it occurs.

You can find descriptions of Darwin's theory in almost any biology text (at least at the college level; sadly, political pressures have caused it often to be watered down for the high school level), but its basic logic was described with particular elegance by biologist Stephen Jay Gould (1941–2002). As Gould put it, Darwin built his model from "two undeniable facts and an inescapable conclusion":

> *Fact 1: Overproduction and competition for survival.* Any localized population of a species has the potential to produce far more offspring than the local environment can support with resources such as food and shelter. This overproduction leads to a competition for survival among the individuals of the population.

> *Fact 2: Individual variation.* Individuals in a population of any species vary in many heritable traits (traits passed from parents to offspring). No two individuals are exactly alike, and some individuals possess traits that make them better able to compete for food and other vital resources.

> *Inescapable conclusion: Unequal reproductive success.* In the struggle for survival, those individuals whose traits best enable them to survive and reproduce will, on average, leave the largest number of offspring that in turn survive to reproduce. Therefore, in any local environment, heritable traits that enhance survival and successful reproduction will become progressively more common in succeeding generations.

It is this unequal reproductive success that Darwin called *natural selection*: Over time, advantageous genetic traits will naturally win out (be "selected") over less advantageous traits because they are more likely to be passed down through many generations. This process explains how species can change in response to their environment—by favoring traits that improve adaptation—and thus is the primary mechanism of evolution. That is, life evolves as natural selection leads over time to evolutionary adaptations that make species better suited to their environments. When the adaptations are significant enough, organisms carrying the adaptations may be so different from their ancestors that they constitute an entirely new species.

Darwin backed his logical claim that evolution proceeds through natural selection by carefully documenting a prodigious amount of evidence. His most famous evidence came from his studies of the unique species of the Galápagos Islands. For example, the islands have 13 distinct finch species ("Darwin's finches"), with different species on different islands and each species adapted to survive in its own peculiar way. Darwin recognized that this made perfect sense when considered in the context of natural selection: Some time in the past, an ancestral pair of finches reached the Galápagos from the mainland (perhaps by being blown off course by winds). Over time, local populations of island finches gradually adapted to different environments, ultimately becoming the distinct species that he observed.

Darwin recognized similar patterns among many other species in the Galápagos and elsewhere in his round-the-world voyage on the HMS *Beagle*, as well as in patterns he saw when comparing fossils of extinct organisms to modern species found in the same regions. He also found strong support for his theory of evolution by looking at examples of *artificial selection*—the selective breeding of domesticated plants or animals by humans. Dogs offer a powerful example: Breeds as different as Rottweilers and Chihuahuas were bred from a common ancestor within just a few thousand years. Darwin recognized that if artificial selection could cause such profound changes in just thousands of years, natural selection could do far more over the millions or billions of years during which it has operated.

Today, we can observe natural selection occurring right before our eyes. In many places on Earth, species have changed in time spans as short as a few decades in response to human-induced environmental changes. On a microbial level, natural selection is what allows a population of bacteria to become resistant to specific antibiotics; those few bacteria that acquire a genetic trait of resistance are the only ones that survive in the presence of the antibiotic. Indeed, bacterial cases of natural selection pose a difficult problem for modern medicine, because bacteria can quickly develop resistance to almost any new drug we produce. As a result, pharmaceutical companies are constantly working to develop new antibiotics as bacteria become resistant to existing ones. Viruses can evolve even faster, which is one reason it has proven so difficult to fight viral diseases such as the common cold, influenza, and AIDS.

All of this observational evidence makes an incredibly strong case for the theory of evolution, but in the past few decades the case has grown far stronger still. Not only do we now observe the results of evolution, but thanks to our modern understanding of DNA, we can explain exactly what takes place on a molecular level. The molecular basis of evolution comes directly from the same imperfect copying of DNA that enables individual variation.

DNA replication proceeds with remarkable speed and accuracy. Some bacteria can copy their complete genomes in a matter of minutes, and copying the complete 3-billion-base sequence in human DNA takes a human cell only a few hours. In terms of accuracy, the copying process generally occurs with less than one error *per billion* bases copied. Nevertheless, errors (mutations) do sometimes occur. For example, the wrong base may occasionally get attached in a base pair, for example linking C to A rather than to G. In other cases, an extra base may be accidentally inserted into a gene, a base may be deleted, or an entire sequence of bases might be duplicated or eliminated. Absorption of ultraviolet light or nuclear radiation or the action of certain chemicals (carcinogens) can also cause mutations to occur in DNA, and once these changes are made they can be copied when the DNA gets copied.

When a daughter cell inherits a mutated DNA molecule, the mutation can affect the functionality of the cell. Many mutations are lethal, in which case the daughter cell does not live to reproduce. However, if the cell survives, the mutation will be copied every time the DNA is replicated. In that case, the mutation represents a permanent change in the cell's hereditary information. If the cell happens to be one that gets passed to the organism's offspring—as is always the case for single-celled organisms and can be the case for animals if the mutation occurs in an egg or sperm cell—the offspring will have a gene that differs from that of the parent. It is this process of mutation, along with the shuffling of genes in sexual reproduction, that leads to variation among individuals in a species (Fact #2 above). Each of us differs slightly from all other humans because we each possess a unique genome with slightly different base sequences.

Mutations therefore provide the molecular basis for evolution.[4] Given that different individuals of a species possess slightly different genes, it is inevitable that some genes will provide advantageous adaptations to the environment. As outlined above, the combination of individual variation and population pressure leads to natural selection, in which the advantageous adaptations are preferentially passed down through the generations. Thus, what was once a random mutation in a single individual can eventually become the "normal" version of the gene for an entire species, thereby explaining how species evolve through time.

Our detailed understanding of how evolution proceeds on a molecular level, coupled with all the other evidence for evolution collected by Darwin and others, puts the theory of evolution by natural selection on a solid foundation. In other words, it is a true *scientific theory*, by which we mean a model that has been carefully checked and tested and that has passed every test yet presented to it. Like any scientific theory, the theory of evolution can never be proven beyond all doubt. But to say it is "only a theory" reveals only ignorance of what it is all about.

[4] Evolution sometimes occurs in an even more dramatic way: In some cases, organisms can transfer entire genes to other organisms, a process called *lateral gene transfer*. This process is one of the primary ways that bacteria gain resistance to antibiotics. We humans have also learned to use this process for our benefit through what we call *genetic engineering*, in which we take a gene from one organism and insert it into another. For example, genetic engineering has allowed us to produce human insulin for diabetic patients: The human gene for insulin is inserted into bacteria, and these bacteria produce insulin that can be extracted and used as medicine. Lateral gene transfer can change a species more rapidly than individual mutations, but mutations are still the underlying basis, since they created the genes in the first place.

EVOLUTION IN THE CLASSROOM

We have covered enough to understand the importance of evolution and how it helps shape our definition of life, which means we should be ready to move on and discuss the implications of these ideas to the search for life on other worlds. But by this point in the book, you probably know I can't step off my soapbox quite so quickly when I'm on a roll, and this topic is way too important to the future of our nation to let it go quite yet. I'm not kidding; you've seen the studies about how America is falling behind in math and science education, and the poor state of education about evolution is a key reason why. After all, if we can't teach our children about the most important and unifying discovery in the history of biology, how can we expect them to learn science at all?

When the opponents of teaching evolution offer their "only a theory" stickers, they are trying to make a distinction between "facts" and "theories." There are some cases in which such a distinction is valid, but in this case it is a false choice, analogous to asking whether gravity is fact or a theory. Gravity is a *fact* in that objects really do fall down and planets really do orbit the Sun, but we use the *theory* of gravity to explain exactly how and why these things occur. The theory of gravity is not presumed to be perfect and indeed has at least one known flaw (its inconsistency with quantum mechanics on very small scales). Moreover, Newton's original theory of gravity is now considered only an approximation to Einstein's improved theory of gravity, which itself will presumably be found to be an approximation to a more complete theory that has not yet been discovered.

The same idea holds for evolution. Nearly all scientists consider evolution to be a fact, because both the fossil record and observations of modern species make clear that living organisms really do change with time. We use the *theory* of evolution to explain how and why these changes occur. The theory of evolution clearly explains the major features of life on Earth, but as with the theory of gravity, scientists still debate the details. For example, there is considerable debate about the rate at which evolution proceeds: Some scientists suspect that evolution is "punctuated," with periods of rapid change followed by long periods in which species remain quite stable, while others suspect that evolution proceeds at a steadier pace. This debate can be quite heated between individual scientists, but it does not change the overall idea that life evolves, and it is a debate that will eventually be settled by evidence. Indeed, we can draw a direct analogy between Darwin's original theory of evolution by natural selection and Newton's original theory of gravity: Just as Newton's theory captured the main features of gravity but has been refined and improved over time, Darwin's theory captured the main features of evolution and has been refined and improved as we've gained a deeper understanding of DNA and relationships among species. And like the theory of gravity, the theory of evolution remains a work in progress. Perhaps someday we'll be able to broaden the theory through the study of comparative evolution, in which we'll explore the similarities and differences among living organisms on multiple worlds. But it is highly unlikely that we'll ever find any fundamental flaw in the basic theory of evolution by natural selection.

Another incorrect claim often made by opponents of teaching evolution is that evolution is not really science. To understand the fallacy in this claim, we need only to look at how evolution stacks up against the three hallmarks of science that I outlined in chapter 2. Evolution clearly satisfies the first hallmark, which states that science seeks explanations for observed phenomena that rely solely on natural causes. It also clearly meets the second, which states that science progresses through the creation and testing of models. For example, the very idea of evolution won out over Aristotle's competing idea of species that never changed, and Darwin's theory won out over Lamarck's earlier model because it explained the observations so much more successfully. The objections from the opponents therefore usually revolve around the third hallmark, which states that a scientific model must make testable predictions that would lead us to revise or abandon the model if the predictions do not agree with observations.

In essence, the opponents claim that evolution is a matter of faith because it does not make testable predictions. *But it does.* For example, the modern theory of evolution, understood on a molecular level, predicts that diseases can and will evolve in response to medicines designed to combat them, a prediction borne out in the rapid way that many diseases acquire drug resistance. It also predicts that genetically similar species should respond to medicines in similar ways, a prediction confirmed by the fact that we can test many medicines in other primates and they do indeed have effects similar to those they have in humans. The theory of evolution also provides a road map that we can use to modify organisms through genetic engineering; in this sense, every genetically engineered grain of rice or corn represents a success of the predictive abilities of the theory of evolution.

In fact, even Darwin's original theory made testable predictions. For natural selection to be possible, Darwin had to assume that living organisms have some way of passing on their heritable traits from parent to offspring. So although he did not predict the existence of DNA per se, his theory clearly predicted that some type of mechanism had to exist to carry the hereditary information. Moreover, now that we understand DNA and its role in heredity, the theory of evolution predicts that closely related species should also be genetically similar, a prediction that has been confirmed in just the past few years by genome sequencing. For example, in the ordering of their base sequences, the DNA of humans and chimpanzees is 98.5 percent identical. Similarly, DNA studies show that primates are all more closely related to one another than to other animals, that animals are all more closely related to one another than they are to plants, that plants and animals together are more closely related to each other than to bacteria, and so on. These relationships are clearly expected according to the theory of evolution, and they would make no sense if we were incorrect about the mechanism by which mutations in DNA make possible natural selection.

Note that none of this makes any statement at all about whether a God or anyone else has had a guiding hand in evolution. Indeed, as you can see in the quotation at the beginning of this chapter, even Darwin himself thought he might be seeing God's hand in creation. Like Darwin and countless other scientists, you are free to believe in your religion and evolution at the same time. You're even free to believe that none of it is true, and that all the scientists who accept it are sadly misguided. But the fact is that these same scientists are using the theory of evolution to advance medicine, agriculture, ecology, and human knowledge of the biological universe. Whether you believe them or not, if you want your kids to grow up able to make similar contributions to our civilization, it is crucial that we teach evolution in school, and teach it well.

THE ENVIRONMENTAL REQUIREMENTS FOR LIFE

Leaving school behind, let's head back out into the universe. Based on the properties of life described in this chapter, we should now be able to come up with a list of the environmental requirements for life. Having such a list should help us to decide which of the many worlds in our solar system or beyond seem like reasonable candidates for harboring life. So without further ado, if we hope to find life on some other world, we expect that it will need:

1. A source of molecules from which to build its own cellular structures and for reproduction.

2. A source of energy to maintain biological order and to fuel the many chemical reactions that occur in life.
3. A liquid medium—most likely liquid water—for transporting the molecules of life.

These three requirements should make sense in light of what we've discussed. The first simply reflects the fact that life creates order out of chaos and reproduces itself. Creating order requires building blocks from which to make it, and making copies requires even more building blocks. Without molecules to serve as such building blocks, it's difficult to see how life would be possible. The need for an energy source is probably similarly self-evident, since the second law of thermodynamics tells us that life would decay into disorder without energy input. The third requirement is a little more subtle, but probably equally universal. If you're going to have lots of chemical reactions with lots of molecular building blocks, you need a way of moving those molecules around both within a living organism and into and out of the organism. Molecules don't move easily through solids, and they tend to be too widely dispersed in gases to be of much use. That means we need a liquid in which the molecular building blocks can be suspended and transported.

Notice that none of these requirements are specific to Earth life; that is, we expect them to be necessary even for alien life that might be biochemically quite different from Earth life, as long as it meets our basic definition of being something that creates order from chaos, reproduces, and evolves. But we can constrain the possibilities more tightly if we focus not only on these most general properties of life, but also on a few specific properties of life on Earth that might also be very common, if not universal.

Let's start with the molecular building blocks. To decide what types of molecules are necessary, we need to think about what life is made of. If you neglect the ubiquitous water that makes up a large part of the mass of all living organisms on Earth, the most important ingredient of life is carbon. We say that life on Earth is *carbon-based* because all the important molecules of life—including proteins, fats, carbohydrates, and DNA—are essentially long chains of carbon atoms attached to various other atoms such as hydrogen, oxygen, and nitrogen. These atoms exist throughout the universe, because they are "star stuff" as we discussed in chapter 1. But to be useful to life, they must be available in a form that allows them to be extracted from the environment. In practice, the availability of carbon is probably the limiting factor.

Where can living organisms obtain their carbon? There are two basic possibilities, illustrated by plants and animals. Plants obtain their carbon directly from the environment, taking in carbon dioxide from air and extracting the carbon for use in building molecules of life. Animals get their carbon by eating plants or other animals. These same basic possibilities hold for all other life on Earth as well. Even the most extreme extremophiles still need a carbon source, which is generally carbon dioxide in the environment. For example, the rock-dwelling endoliths obtain carbon from tiny bubbles of carbon dioxide that have filtered down from the surface, while hyperthermophiles like *P. fumarii* and Strain 121 get carbon from carbon dioxide dissolved in the ocean.

Of course, when we consider the possibility of extraterrestrial life, it's natural to wonder whether it might be based on an element besides carbon. In truth, we cannot say for sure whether other elements would work. However, carbon has something going for it that seems quite useful to life: It has the ability to form chemical bonds to up to four other atoms at once, and it sometimes forms stronger *double bonds* to other carbon atoms. Given the importance of these bonding properties of carbon, we might expect that any other elemental basis for life would have to have the same capabilities. Among the elements common on Earth's surface—and likely to be common on other planets—silicon is the only element besides carbon that can bond to four atoms at once. As a result, science fiction writers have often speculated about finding silicon-based life on other worlds.

Unfortunately for science fiction, silicon has at least three strikes against it as a basis for life. First and most important, the bonds formed by silicon are significantly weaker than equivalent bonds formed by carbon. As a result, complex molecules based on silicon are more fragile than those based on carbon—perhaps too fragile to form the structural components of living cells. Second, unlike carbon, silicon does not normally form double bonds; this fact limits the range of chemical reactions that silicon-based molecules can engage in as well as the variety of molecular structures that can form. Third, while carbon can be easily extracted from the environment in the form of gaseous carbon dioxide, silicon is generally found only in solid forms (such as silicon dioxide, which makes up quartz and several other minerals); as a result, there would be no easy way for life to extract silicon from the environment. Given the three strikes against silicon, most scientists consider it unlikely that life can be silicon-based. Moreover, observational evidence on Earth also argues against silicon: Silicon is about 1,000 times as abundant as carbon in Earth's crust, so the fact that life here is carbon-based despite the greater abundance of silicon suggests that carbon will always win out over silicon as a basis for life.

A few other elements have also been suggested as possibilities for replacing carbon on other worlds, but many scientists believe carbon's natural advantages will still win out. We have found carbon-based molecules even in space (as identified in meteorites and interstellar clouds), suggesting that carbon chemistry is so easy and so common that even if life with another basis were possible, carbon-based life probably would arise first and then reproduce so successfully that it would crowd out the possibility of any other type of life. We cannot be certain that carbon is a general requirement for life, but it seems reasonable to at least start the search for life elsewhere by focusing on places where carbon should be available.

Let's turn next to the energy requirement. For energy, life on Earth uses one of three basic sources. The first source is sunlight: Plants and many microbes obtain energy directly from sunlight through the process of photosynthesis. The second is food: Animals (and many microbes) obtain energy by breaking down molecules that were built by plants or other organisms that they consume. The third and perhaps most amazing energy source is chemical reactions in the surrounding environment. This is the energy source for many extremophiles that need neither organic food nor sunlight to survive. For example, the archaea known as *Sulfolobus* live in volcanic hot springs and obtain energy from chemical reactions involving sulfur compounds, while *P. fumarii* supplements the nutrition it obtains from dissolved carbon dioxide with energy from chemical reactions between hot water and the rock walls of the black smokers in which it lives. Notice that this type of chemical energy can be found almost anywhere that even tiny amounts of liquid water are moving among minerals in rock, which leads us to the third requirement, the liquid medium.

The need for a liquid medium of some type seems hard to get around, but does it really need to be water, or might other liquids work as well? Again, we really don't know, but there are several constraints to consider. For example, a substance that might fulfill the roles of water must, like water, be fairly common. On Earth, the only common liquid besides water is molten rock, which is so hot that it's difficult to imagine life surviving within it. However, several other common substances can take liquid form on colder worlds. For example, liquid methane apparently can exist on the surface of Saturn's moon Titan, and liquid ammonia—or an ammonia–water mixture—may be found beneath the surfaces of numerous moons of the outer solar system.

The primary problem with these "other" liquids is that they exist in liquid form only at extremely cold temperatures: Liquid ammonia boils away at a temperature of about –28°F, while liquid methane exists only at temperatures below about –260°F. Chemical reactions generally proceed much more slowly at these cold temperatures than at the warmer temperatures at which water is liquid, and some biologists argue that the pace of chemical reactions in these liquids would be too slow to support life (though other biologists disagree, arguing that slow reactions are not necessarily an impediment to life). And even if the pace of chemical reactions is not enough to rule out life, water has at least two other properties that seem potentially important to life and that these other liquids lack: Water ice is less dense than liquid water, which is why ice floats, and water molecules have electrical charge separation ("polarity") that enables them to form types of chemical bonds that the other liquids cannot. The fact that ice floats is useful to life on Earth because it allows ice to insulate liquid water beneath it in ponds and lakes, leaving a place where life can survive in winter. The chemical bonding properties of water come into play in a great variety of the biochemical processes that occur in living organisms on Earth. Overall, while it's marginally possible that cold liquids such as methane or ammonia might support life, it seems far more likely that any life we someday discover will be using water as its liquid medium.

BEYOND UFOS

I began this chapter with the idea that short of an alien landing or a clear SETI detection, we're going to have to rely on science to determine whether life exists elsewhere in the universe. With the vast number of worlds that exist out there, this means we need a way of intelligently planning and organizing a search. The first step in such a search is to understand what we are searching for, which is why we spent some time discussing the basic properties and definition of life. The next step is to decide where to look, and we've found that our understanding of life gives us some reasonable guidance: While it's always possible that we may be too limited in our thinking about life, the best way to start seems to be by looking for worlds that have a carbon source, available energy, and a liquid medium that is most likely to work if it is liquid water.

In fact, the availability of carbon and energy are probably covered automatically whenever the liquid requirement is met, especially if it is liquid water. As we've discussed, all the important chemical elements for life should be widely available on most worlds, and the only real question is whether they exist in a form that allows them to be extracted by life. Whenever liquid water is present, it is likely to have dissolved minerals and gases that mean the elements would indeed be in an extractable form. Similarly, because a rock-water interface can almost always support energy-releasing chemical reactions, the presence of liquid water also means at least some available energy for life. Putting these ideas together, we can consolidate the requirements for life into a single "litmus test" for the habitability of another world: *A world can be habitable—meaning that it offers the potential to support life—only if it has a liquid medium, probably meaning liquid water.*

This litmus test certainly narrows the possibilities for abodes of life, but not as much as you might at first guess. The wide variety of habitats in which we find both liquid water and life on Earth, including the deep ocean and rocks buried deep underground, tells us that habitability requires only the presence of a liquid *somewhere,* not necessarily on the surface. As we'll discuss further in chapter 7, a large fraction of the worlds in our solar system have probably met this condition at least some time in the past, and several may still meet it—including Mars, Europa, and Titan. The possibilities are surely much greater if we look to other star systems.

And this idea brings us to our next step. Based on our discussions in this chapter, it seems reasonable to imagine that if we *put* life on various other worlds such as Mars or Europa, it might actually survive and perhaps even thrive—which is one reason that scientists take great care not to contaminate other worlds with spacecraft that we send to them. But could these worlds already have their own, indigenous life? More to the point, should we expect to find life on the numerous but widely separated worlds found around the myriad other stars in our galaxy and universe? The answer to these questions depends not just on having conditions for the survival of life, but on how easily life can arise on a world that offers habitable conditions, which will therefore be the topic with which I'll start the next chapter.

Plate 1. This Hubble Space Telescope photograph, called the Hubble Ultra Deep Field, shows galaxies visible to the Hubble Space Telescope in a tiny piece of the sky no larger than a grain of sand held at arm's length.

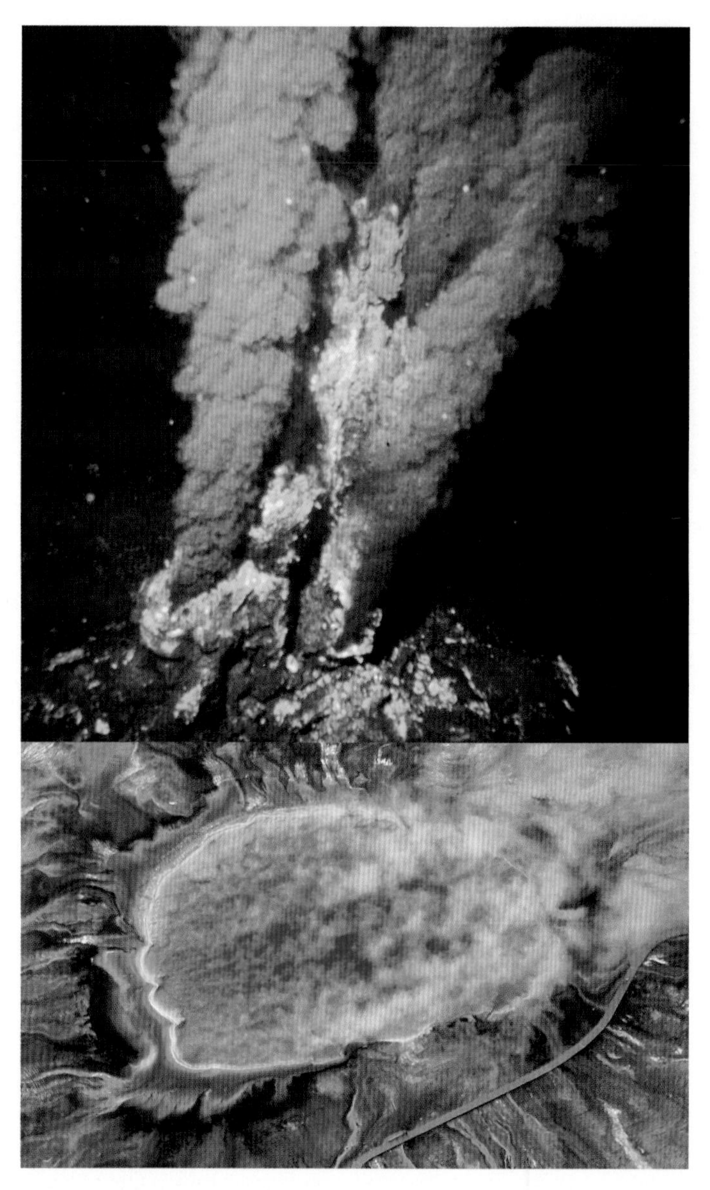

Plate 2. Microbial life thrives here. (Top) A volcanic vent on the seafloor, known as a black smoker, spewing out extremely hot, mineral-rich water. Organisms like P. *fumarii* and Strain 121 survive here in water above the normal boiling temperature. (Photo by P. Rona; courtesy of OAR/National Undersea Research Program (NURP); NOAA) (Bottom) Grand Prismatic Spring, a hot spring in Yellowstone National Park; the walkway winding along the lower right provides scale. The different colors in the water are from different bacteria that survive in water of different temperatures. (National Park Service photo by Jim Peaco.)

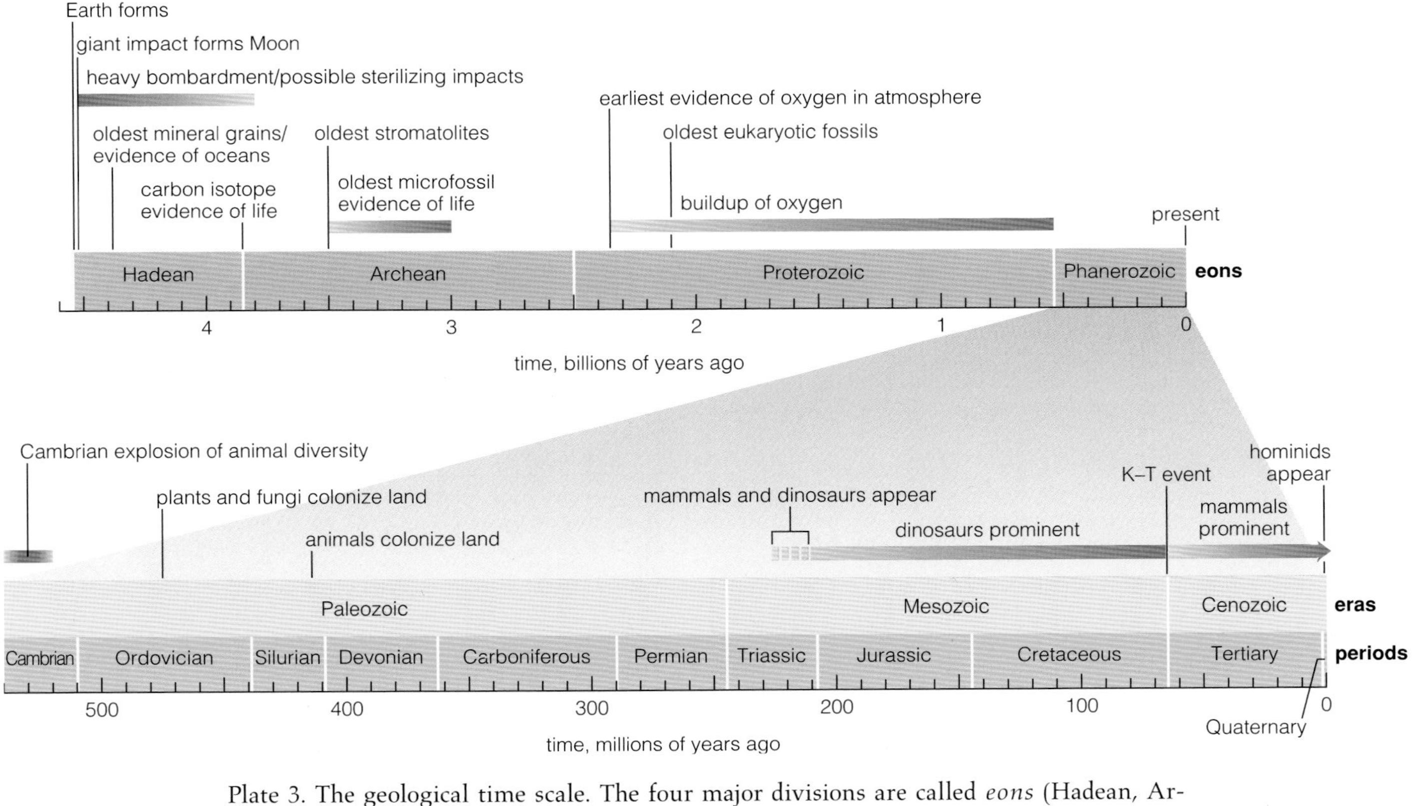

Plate 3. The geological time scale. The four major divisions are called *eons* (Hadean, Archean, Proterozoic, Phanerozoic). The Phanerozoic eon, shown expanded on the lower zoom-out, is subdivided as indicated into *eras* and *periods*. (Illustration courtesy of Addison Wesley, an imprint of Pearson Education.)

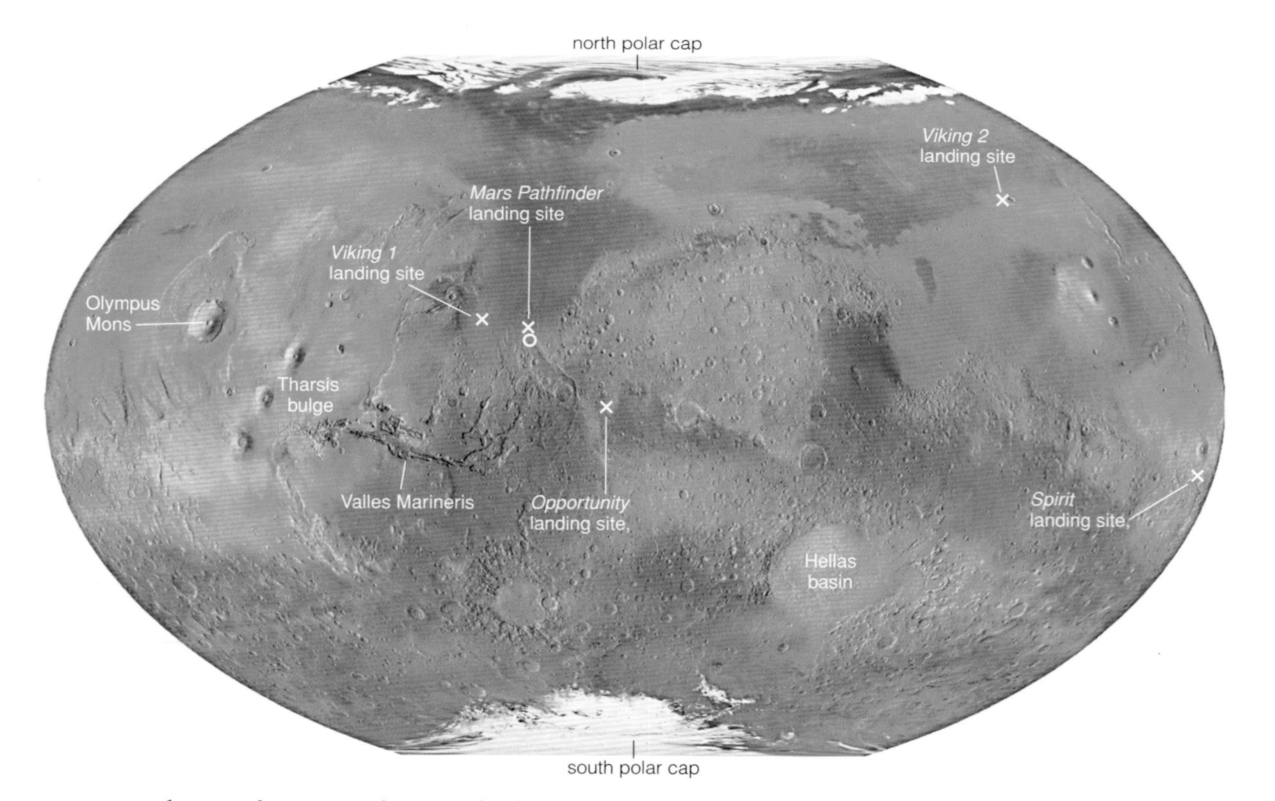

Plate 4. This image showing the full surface of Mars is a composite made by combining more than 1,000 images with more than 200 million altitude measurements from the Mars Global Surveyor mission.

Plate 5. This sequence zooms in on a rock outcropping near the *Opportunity* rover's landing site. The outcrop stands about knee-high. The close-up shows a piece of the rock a little more than an inch across. The layered structure, the odd indentations, and the small spheres (nicknamed "blueberries") all support the idea that the rock formed from sediments in standing water. (Courtesy NASA/JPL-Caltech.)

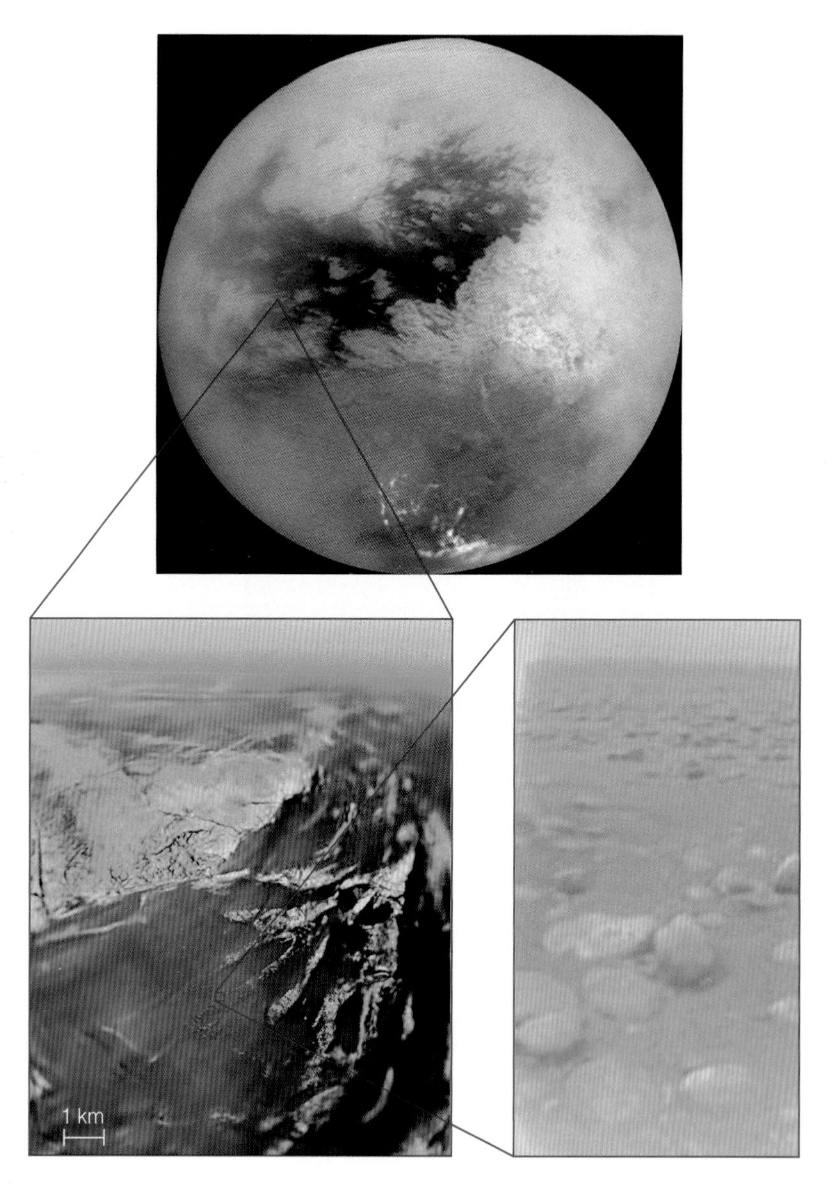

Plate 6. This sequence zooms in on the *Huygens* landing site on Titan. (Top) a global view taken by the Cassini orbiter. (Left) An aerial mosaic of images taken by the *Huygens* probe during its descent. (Right) A surface view taken by the probe after landing; the "rocks," which are a few inches across, are presumably made of ices. Keep in mind that you are looking at the surface of a world nearly a billion miles away. (Courtesy NASA/ESA)

Plate 7. Active Enceladus. (Top) Enceladus in daylight. The blue "tiger stripes" are regions of fresh ice that must have recently emerged from below. The colors are exaggerated; the image is a composite made at ultraviolet, visible, and infrared wavelengths. (Bottom) Enceladus backlit by the Sun. Fountains of ice particles (and water vapor) are clearly visible as they spray out to the lower left.

Plate 8a. The partially completed Allen Telescope Array in Hat Creek, California. The array is a joint project of the SETI Institute and the University of California at Berkeley. (Photo by Seth Shostak)

Plate 8b. Given that the total number of stars in our universe is about as great as the total number of grains of sand on all Earth's beaches, can it really be the case that we are alone? It's possible, but it seems unlikely. (Photo by the author)

5

GETTING LIFE STARTED

> The more I examine the universe and study the details of its architecture, the more evidence I find that the universe in some sense must have known we were coming.
>
> —*Freeman Dyson, in* Disturbing the Universe

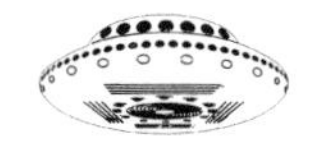

As a scientist who speaks frequently to audiences ranging from school children to college students, teachers, and the general public, I've learned not to be too surprised when audience expectations differ from my own. For one thing, I never expect an audience to be particularly large, unless the professors at a local college are offering extra credit for their students to attend (extra credit works like a charm at filling lecture halls!), which is why I'm always flattered by the two people who show up a half hour early because they're worried about finding seats. My talk topics can also lead to mismatched expectations. I was at the College of Charleston a few years back, where the astronomy faculty had graciously arranged for me to offer a public talk based on my prior book for the general public. They'd warned me that few students would be able to attend, and that other public talks on astronomy rarely drew more than a few dozen people. So imagine our surprise when the room with 75 chairs was packed standing room only with crowds overflowing out the door! There was a simple explanation, of course: The local newspaper had misprinted my credentials, saying that I was an *astrologer* rather than an *astronomer.* To their credit, not a single person walked out when I explained that I'd have nothing at all to say about horoscopes, and was instead planning to limit my remarks to topics such as the scale of space and time and the birth and fate of the universe. Did I disappoint them? I suppose I'll never know.

These days, most of my talks for general audiences are either talks for kids based on my children's books or talks for grownups based on this book. The kids tend to come without any major expectations, aside from hoping that I won't be boring. Still, at least for the lower grade levels, I know they're usually a little disappointed to learn that my stories about Max the Rottweiler going to the Moon and Mars and Jupiter are not really true, and that I've never even been to those places myself. I guess they thought they were going to meet a real astronaut. Fortunately, a few cute photos of Max and our new dog Cosmo usually do the trick at overcoming this disappointment. For the grownups at my Beyond UFOs talks, I've found that as much as half the audience may have come because they thought I was going to talk *about* UFOs rather than moving beyond them. Indeed, that's one of the reasons I now emphasize the word *search* in my subtitle, hoping to make clear that I don't plan to spend much time on flying saucers.

Even then, however, the most common audience expectation is that I'm going to talk primarily about SETI, the search not just for extraterrestrial life of any kind but for extraterrestrial *intelligence*. For most people, SETI is the public face of astrobiology, holding out the potential promise of introducing us to advanced new friends in the cosmos. So most people are usually surprised when I explain that, while SETI is an important part of the search for life in the universe, it is not by any means the only focus of scientific research. It isn't even the sole focus of research at the SETI Institute in Mountain View, California; as the SETI scientists well know, there'd be little point in searching for intelligence if we weren't at least moderately confident that there are many worlds with microbial life. But should we really be so confident?

Way back in chapter 1, I argued that the biological context gives us three reasons for thinking that life might indeed be common (see figure 1.2). So far, we've come back to touch on only one of those reasons: the idea that the great diversity of life on Earth suggests that life can survive in a much wider range of conditions than we thought possible just a few decades ago. As we discussed in the prior chapter, some of the extremophiles discovered in recent years here on Earth seem like they might survive the conditions found even on other worlds of our own solar system, not to mention countless worlds beyond. But the potential for life to survive on a distant world is not enough to make life likely. For that, we need reason to think that life could get started in the first place, and that brings us to the other two reasons we discussed in chapter 1: that the building blocks of life are common and that life arose quickly here on Earth. Together, these two ideas would give us reason to think that life might arise with similar speed elsewhere. But what makes us think these ideas are true? In this chapter, I'll try to explain the evidence behind them. Then, once we understand why it seems reasonable to imagine an independent origin of life on other worlds, we'll explore the question of whether it also is likely for life to go from the microbial to the more complex, perhaps ultimately giving rise to civilizations.

THE ORIGIN OF LIFE ON EARTH

One of the hottest areas of research in astrobiology concerns not the search for actual life elsewhere, but instead the search to understand life's origins right here on Earth. There's a simple reason for this: Earth is the only place we know of that without doubt has life, so anything we learn about how life might have started here should give us deep insights into the possibilities for elsewhere. We can break the quest to understand life's origins into three related questions: *When, where,* and *how* did life arise on Earth? We do not yet know precise answers to any of these questions, and it's possible that we never will; after all, we have only the limited clues left in rocks, fossils, and the life that survives to this day to reconstruct what happened in the distant past. Still, like detectives at a crime scene, we can use these clues to construct a timeline and plausible scenarios of what must have occurred. Let's start with the timeline, which gives us clues about when life first arose.

In many places around the world, you can see canyon walls layered with Earth's history. The rock layers of the Grand Canyon, for example, record a fairly continuous history of the past 500 million years. Geologists can date the layers with the technique of radiometric dating (see chapter 3), and the fossils within the layers then tell us about the history of life on Earth. By combining data from many different sites around the world, scientists have put together a timeline of Earth's history, usually called the *geological time scale* and shown in color plate 3.

One of the first things that will jump out at you as you study the geological time scale is that we know a lot more about recent times than older times. There are at least three reasons for this fact. First, older rocks are rarer than younger ones, because many rocks are destroyed over time. Second, the demarcations between periods, eras, and eons are based on changes in the fossils and sediments that we find for each time range, and fossils also become rarer as we look back in time. Third, life itself has changed with time; prior to the start of the Cambrian explosion about 540 million years ago (which marks the boundary between the Proterozoic and Phanerozoic eons), most life on Earth was microscopic, and fossils of microscopic life are much more difficult to find than, say, fossils of dinosaur bones. These three difficulties mean the fossil record will never be fully "complete," since there will always be life forms that existed but that left no fossils that we have yet discovered. Creationists sometimes claim this to be a flaw in our reading of the past, but it is really no different than the crime scene analogy I described above. The past always leaves an incomplete record, but with careful study and enough clues we can still determine what happened.

Incidentally, since I'm relatively new to geology myself (having been trained as an astronomer), I was curious about where all the time periods on the geological scale got their names. So I looked it up and now I can tell you, too. The names of the four *eons* have Greek roots. The Phanerozoic eon, which extends from the present back to about 540 million years ago, takes its name from the Greek for "visible life" because it is marked by the presence of fossils visible to the naked eye. The Proterozoic eon, which extends from 540 million to about 2.5 billion years ago, means the eon of "earlier life" because it shows fossils of single-celled organisms that lived before the Phanerozoic. The Archean eon extends from 2.5 to about 3.85 billion years ago and is named for "ancient life"; it got this name after the discovery of fossils from the first half of Earth's history. The Hadean eon, which dates from 3.85 billion years ago back to Earth's birth some 4.55 billion years ago, got its name because scientists used to presume that the early Earth would have had "hellish" conditions (Hades was the Greek mythological name for the underworld); however, recent evidence from very old mineral grains suggests that the Hadean may not have been quite that bad, and that Earth may have had oceans and continents as early as 4.4 to 4.5 billion years ago.

The fact that the geological record is much richer for more recent times is reflected in the more detailed naming system used for these times. The Phanerozoic eon is subdivided into three major *eras*: the Paleozoic, Mesozoic, and Cenozoic. These names also have Greek roots and mean, respectively, "old life," "middle life," and "recent life." The three eras are further subdivided into *periods*. The periods do not follow any consistent naming scheme. For example, the Cambrian period gets its name from the Roman name for Wales (in Great Britain), the Jurassic period gets its name from rocks found in the Jura mountains of Europe, and the Tertiary period simply means "third" period. The recent geologic periods are even further subdivided into *epochs* and *ages*, but these are not shown on the timeline in color plate 3.

The fossil record does not tell us exactly when life arose, because there is no way to know if we have found fossil evidence of the earliest life forms.

However, if we have a fossil of a certain age, then life must be at least that old. So how old is the oldest fossil evidence? Alas, it's not an easy question to answer. Very old rocks are not only rare, but they are also likely to have been distorted (metamorphosed) by heat or pressure over time, thereby breaking up any microscopic fossils that they might contain. As a result, we must study very old rocks with extreme care if we seek evidence of life, and the required measurements are difficult enough that they often engender scientific controversy. But putting aside the details, I'll give you the bottom line here.

The oldest, well-accepted evidence for fossil life comes from rocks called *stromatolites* that date to ages as old as about 3.5 billion years. In size, shape, and interior structure, ancient stromatolites look virtually identical to sections of mats formed today by colonies of microbes sometimes called "living stromatolites" that are found in places including the World Heritage site at Shark Bay, Western Australia. Living stromatolites contain layers of sediment intermixed with different types of microbes. Microbes near the top generate energy through photosynthesis, and those beneath use organic compounds left as waste products by the photosynthetic microbes. The living stromatolites grow in size as sediments are deposited over them, forcing the microbes to migrate upward in order to remain at the depths to which they are adapted. The similarity of structure between the ancient stromatolites and the modern-day mats suggests a similar origin, and recent studies (by Abigail Allwood, now at NASA's Jet Propulsion Laboratory, and her colleagues) offer further support for the conclusion that stromatolites were made by ancient microbes. Moreover, if the microbes that made the stromatolites are like the microbes in the living mats today, then the implication is that at least some of these ancient microbes produced energy by photosynthesis. Because photosynthesis is a fairly sophisticated metabolic process, we presume that it must have taken at least a moderately long time for this process to evolve in living organisms. In other words, the existence of 3.5-billion-year-old stromatolites suggests that life itself had already been around for some time when they formed.

Other evidence for life long ago comes from microfossils—structures that appear to be fossils of individual microorganisms. Like the oldest stromatolites, the oldest microfossils date to about 3.5 billion years ago, though there's some controversy over whether these particular structures are truly fossils or if they were created by mineral processes (rather than biological processes) and only bear a coincidental resemblance to fossilized cells. Further study will eventually settle the question of whether these structures are truly fossilized cells, but in the meantime they hardly change the more general picture, because other microfossils are nearly as old and much less controversial. For example, microfossils have been found at two sites in southern Africa in rocks that date to between 3.2 and 3.5 billion years old. By the time you get to rocks just a little younger than 3 billion years old, microfossil evidence seems abundant and clear.

Together, the stromatolites and microfossils make a strong case for life already being abundant and fairly sophisticated (at least for microbes) by 3.5 billion years ago. Another line of evidence takes us back even farther. As shown in figure 3.2, carbon comes in three different forms, or isotopes, known as carbon-12, carbon-13, and carbon-14. The latter is radioactive, and its relatively short half-life of 5,700 years ensures that none at all can remain from billions of years ago. The other two isotopes of carbon are stable, and in inorganic material they are always found in a characteristic ratio of about one carbon-13 atom for every 89 carbon-12 atoms. The ratio changes in living organisms, however, because they incorporate carbon-12 atoms into cellular molecules slightly more easily than they do carbon-13 atoms. As a result, living organisms—and fossils of living organisms—always show a slightly lower fraction of carbon-13 atoms than that found in inorganic material. On the island of Akilia off the coast of Greenland, University of Colorado geologist Stephen Mojzsis has found this lower ratio of carbon-13 to carbon-12 in rocks that are more than 3.85 billion years old, seemingly making a clear case that life already existed by that time (figure 5.1). A few scientists have questioned Mojzsis's interpretation of the data, but because he works at my home institution I had a chance to sit down with him and learn for myself some of the details behind the conflicting claims. He makes a strong case, and personally I'll be very surprised if his claims do not hold up under further scrutiny. Moreover, because the geological record is so sparse for such early times, the carbon isotope evidence would imply not only that life existed at that time, but that life must already have been widespread on Earth.

Given that Earth formed about 4.55 billion years ago, the evidence of widespread life prior to 3.85 billion years ago suggests that the origin of life occurred within the first 0.7 billion, or 700 million, years of our planet's history. Geologically speaking, this is quite early in Earth's history, and it already demonstrates that life took hold fairly quickly on our planet. However, if you look again at the timeline in color plate 3, you'll see something that makes it likely that life arose much faster still.

Notice the segment on the timeline marked "heavy bombardment" extending from Earth's formation to about 3.9 billion years ago. Radiometric dating of Moon rocks shows that most of the Moon's visible impact craters

Figure 5.1. Dr. Stephen Mojzsis, standing with the rock formation off the Greenland coast that contains evidence that life on Earth was already widespread more than 3.85 billion years ago. (Photo courtesy of Stephen Mojzsis.)

must have formed during this early period of the solar system's history. This is not surprising: According to our modern theory of solar system formation, the planets were built as larger and larger chunks of rock (sometimes mixed with metal or ice), or *planetesimals*, collided with one another. When a colliding planetesimal stuck to a growing planet, the planet got larger, increasing its gravity and allowing it to draw in even more planetesimals. Even after the planets had reached essentially their current sizes, there must still have been many planetesimals floating around; some of them still remain today, as the objects we call asteroids and comets. Those planetesimals that had orbits intersecting the orbits of the planets were doomed to eventual collisions, and most of those collisions must have occurred early in the solar system's history, when the number of planetesimals was still large. In other words, the heavy bombardment was the period of time during which impacts were most common, and the evidence from the Moon tells us that this period ended by about 3.9 billion years ago.

Some of the planetesimals were quite big. Indeed, although I won't go into the details in this book, we have good reason to think that the Moon itself was created when a planetesimal the size of the planet Mars struck the young Earth within just 20 to 30 million years after Earth's formation. This "giant impact" (also shown on the timeline) is thought to have blasted rock from Earth's outer layers into space, where some of it settled into Earth orbit and then was collected together by gravity to make the Moon.

Once the Moon formed, it became a record of the continuing impacts, not only telling us when the heavy bombardment occurred but also telling us about the sizes of the impacting objects from the sizes of the craters they left. Because the Apollo missions visited and brought back rocks from only six sites on the Moon, we have only incomplete data about lunar cratering. Nevertheless, these data point to two key ideas: First, while there were no more Mars-size impacts (fortunately!), the Moon continued to be pelted by objects tens of miles to a couple hundred miles across. Second, some of the largest impacts occurred as the heavy bombardment was ending, marking what many scientists now call the "late heavy bombardment." These large impacts created the smooth, lunar *maria* that you can see easily with a pair of binoculars.

Because the heavy bombardment was a phenomenon of the solar system, it cannot have been unique to the Moon. This explains why we see craters on so many other planets and moons. Earth, too, must have been frequently scarred by large impacts during the heavy bombardment. In fact, Earth should have been hit even more than the Moon, because our planet presents a bigger target and Earth's stronger gravity would have drawn in more objects and accelerated them to higher speeds by the time they hit the ground. The only reason we don't see the craters from these impacts on Earth is that they were erased long ago by volcanic eruptions, erosion, and other geological processes that occur here but not on the Moon (for reasons we'll discuss in the next chapter).

What does all this have to do with the origin of life? Calculations suggest that some of the larger impacts would have had devastating effects on life. For example, the impact of an object larger than about 225 miles across would have released enough energy to completely vaporize the oceans and raise the global surface temperature to more than 3,000°F. Such an impact probably would have sterilized our planet, wiping out any life that existed when it occurred. Somewhat smaller impacts would have vaporized all but the deepest ocean water, killing off any life that wasn't either living near the ocean bottom or in rock deep underground.

Because Earth does not retain a cratering record like the Moon's, we have no way to know the precise size or timing of the large impacts on the early Earth. Nevertheless, the example of the Moon tells us that at least a few impacts large enough to sterilize much or all of our planet should have occurred during the heavy bombardment, quite likely continuing through the late heavy bombardment. In other words, *if* life existed on Earth before the end of the heavy bombardment about 3.9 billion years ago, there's a good chance that it would have either been wiped out completely or at least wiped out to the point that only deep ocean and underground life could survive. If you combine this idea with the fact that we have evidence of life before 3.85 billion years ago—or just 50 million years or so after the end of the heavy bombardment—the implication should be clear: While we *know* that life got started on Earth in no more than several hundred million to a billion years, there's a very good chance that it got started in a far shorter time, perhaps just tens of millions of years or less.

By itself, this rapid origin of life proves nothing about life elsewhere, since it is always possible that Earth was just the lucky beneficiary of a highly improbable event. However, if we assume that what happened here would be typical of what might happen elsewhere, then the early origin of life is profoundly important: It suggests that we could expect life to arise similarly rapidly on any other world with similar conditions. And the best way to determine whether we should think that what happened here would be "typical" is to continue on to the *where* and *how* questions about the origin of life on Earth.

WHERE LIFE AROSE ON EARTH

We are even less likely to learn precisely where or how life arose on Earth than to learn exactly when it happened, but we still have clues that can give us good ideas about the most probable scenarios. Perhaps surprisingly, the most important clues for the where question come from genetic studies of modern-day microbes.

All known life on Earth shares a number of striking biochemical similarities that, as far as we know, could easily have been different. For example, all life uses DNA and the same basic genetic code, but we know of no reason why some other molecule or some other code could not have worked equally well. Similarly, all life on Earth stores and releases cellular energy with the same molecule, adenosine triphosphate (ATP), and all life on Earth builds proteins from the same set of about 20 amino acids, even though dozens of other amino acids exist in nature. These commonalities cannot be coincidental, so unless there are some unknown physical laws that make these the only possible molecules of life, they are telling us that all known life on Earth evolved from a common ancestor that used these particular molecules.

The idea that all life on Earth shares a common ancestry gives us a remarkable power: It means that we should be able to reconstruct the evolutionary history of much of life on Earth simply by comparing the genomes of different organisms that are living today. You can understand the idea by thinking about the DNA of the organism that long ago became the common ancestor of all life today. Mutations created variations on this DNA, and each new species therefore had slightly different DNA sequences than did the older species from which it evolved. Over millions and billions of years, continuing evolution led to new species with DNA molecules increasingly different from the DNA of the common ancestor. But, always, the new molecules were built by changes to the older ones so that, in principle, the changes are traceable in the precise base sequences of living organisms. All you need to do is compare the DNA sequences among similar genes in different species: The more closely two species are genetically related, the more recently they must have diverged from a shared ancestor.

This technique of genetic comparison is the same one that I talked about in the prior chapter in describing how biologists put together the tree of life (see figure 4.2). The idea of common ancestry just adds a new layer of interpretation: In addition to telling us how closely species are related, the tree also tells us how far back in time their genes evolved. As we get closer to the "root" of the tree, we must be looking at organisms that have genes more similar to those of the common ancestor than do species on the outer branches. In other words, our best guess of what the common ancestor must have been like some 4 billion years ago is that it probably looked somewhat like the modern-day organisms nearest to the root of the tree. And what organisms are these? Drum roll, please. . . . The organisms closest to the root of the tree of life, and hence most likely to resemble the common ancestor of life, are hyperthermophiles living in the hot water near the deep-sea volcanic vents that we call black smokers.

If you think about it, the idea that life might have first arisen near deep-sea vents makes perfect sense. Early life could not survive on land, because the lack of oxygen in Earth's early atmosphere meant no ozone layer (ozone is a form of molecular oxygen with three oxygen atoms per molecule rather than the two that are in the oxygen we breathe), and no ozone meant no protection from lethal ultraviolet radiation from the Sun. You wouldn't have had to go far underwater to gain protection, but being near the surface would have been of little use: As I noted earlier, photosynthesis is a fairly complex process that probably could not have existed in the earliest life forms, and without photosynthesis they would have had no way to take advantage of the energy of sunlight. Deep-sea vents offered energy of a different sort, through chemical reactions between the hot water and minerals. Moreover, life near deep-sea vents would have been protected from all but the largest impacts of the heavy bombardment.[1] While we may never know for sure, deep-sea vents look like a good bet for the site at which life first took hold on Earth.

HOW LIFE AROSE ON EARTH

Darwin's theory of evolution explains how a single common ancestor could have evolved over 4 billion years into all the diversity of life on Earth today, but it does not tell us how the common ancestor itself came to be. The fossil record is of no help either, because it's unlikely that any evidence of the processes would remain from such a distant time in Earth's past, and even if it did we have no idea how we'd recognize it. Moreover, the nature of modern life seems to pose a chicken and egg question; for example, DNA is copied with the help of proteins that are made according to instructions in the DNA, making it difficult to see how this interdependency originated. Nevertheless, over the past few decades laboratory experiments have given us insights into the types of chemical processes that likely occurred on the early Earth. While these experiments have not yet told us precisely how life first arose, and it's possible that they never will, they give us good reason to think that life may have started through natural, chemical processes. To see why, we need to start by asking how the chemical building blocks of life came to be.

Life today is based on the chemistry of a wide variety of organic molecules, making it logical to assume that the first life was somehow assembled from pre-existing organic molecules on the early Earth. But how would those molecules have come to exist? Turns out, it was probably easy. Since about the 1950s, scientists have been conducting laboratory experiments in which they attempt to recreate the chemical conditions that existed on the early Earth, and then see what happens under those conditions. The only major uncertainty in these experiments is in deciding what composition to assume for Earth's early atmosphere, but the uncertainties today are much

[1] This protection from large impacts means we can't necessarily conclude that the deep-sea vents are the likely site of life's origin; we can only say that this is where the life that evolved to the present day survived after the last major impact. Still, given the other advantages of the deep-sea-vent environment, it seems reasonable to suppose that life actually arose in these locales.

less than they were a few decades ago. The results can vary dramatically depending on the precise assumptions, but the general conclusion is this: The major molecules of life—including amino acids, nucleic acids, carbohydrates, and lipids—form naturally under conditions that prevailed in at least some places in the early oceans of Earth. The only real question is the abundance of these molecules, which remains a topic of great scientific debate. However, there's one place where the experiments tell us that the organic molecules should have formed with particular ease: in the warm mix of water, carbon dioxide, and minerals that would have surrounded the deep-sea vents that we've identified as the most likely sites for an origin of life.

In addition, Earth should have received abundant organic molecules from space. Studies of asteroids and comets show they contain lots of organic molecules, including complex molecules such as amino acids, and these molecules would therefore have come to Earth with every impact of the heavy bombardment. Theoretical studies show that organic material also would have been produced as solar ultraviolet light irradiated grains of interplanetary dust floating throughout the solar system, and these grains should have rained down on the young Earth as well. Overall, the combination of local chemical reactions and organic molecules from space means that our planet had all the necessary building blocks from which to assemble life. In essence, the young Earth had at least some places, including the deep-sea vents, that served as giant, natural laboratories for organic chemistry. Before I go on, notice that this idea fills in the upper left bubble from figure 1.2. That is, we've found that organic material forms easily and naturally, both under the conditions of the young Earth and in space, giving us reason to believe that many other worlds should also be loaded with the building blocks of life.

Of course, to paraphrase the late Carl Sagan, these organic building blocks represent only the notes of the music of life, not the music itself. Viewed in terms of probability, the likelihood of a set of simple building blocks ramming themselves together to form a complete living organism is at least as small as that of letting monkeys loose in a roomful of musical instruments and hearing Beethoven's Ninth Symphony. It simply wouldn't happen, even if the experiment were repeated over and over again for billions of years. If we are going to find a natural explanation for the origin of life, it must include a few intermediate steps—each involving a chemical pathway with a relatively high probability of occurring—that eased the transition from chemistry to biology. Here, too, recent experiments give us at least some reason to think that going from building blocks to actual life was also much easier than we might have guessed.

Since we've defined life as a chemical system that can reproduce and evolve, the key to getting life started would seem to lie with getting a molecule that can serve the hereditary function of DNA. As I've already noted, DNA itself poses a chicken and egg problem: DNA seems far too complex, and its replication far too intertwined with proteins and other molecules, to have been the genetic material of the first living organisms. We are therefore looking for a molecule that is simpler than DNA but still capable of making fairly accurate copies of itself. The most obvious candidate is the molecule called RNA, which like DNA has long strings of chemical bases that can encode genetic information and serve as a template for replication but is much simpler in its overall chemical structure and has only a single strand (rather than the double-stranded helices of DNA).

RNA is indeed capable of replication, and is actually used as the genetic material in some viruses (including HIV). However, the replication of RNA, like that of DNA, is intertwined with numerous other molecules, which for a while made it seem subject to the same chicken and egg dilemma. A way around this dilemma was discovered in the early 1980s by Thomas Cech and his colleagues at the University of Colorado, Boulder. They found that RNA can catalyze biochemical reactions in much the same way enzymes do (work for which Cech shared the Nobel Prize in 1989). We now know that RNA molecules play this type of catalytic role in many cellular functions, and we call such RNA catalysts *ribozymes* (by analogy to enzymes). Follow-up work has shown that some RNA molecules can at least partially catalyze their own replication. These discoveries have led biologists to envision that modern, DNA-based life may have arisen from an earlier *RNA world,* in which RNA molecules served both as genes and as chemical catalysts for copying and expressing those genes.

How might an RNA world have gotten started? The first requirement would have been the spontaneous production of self-replicating strands of RNA. Even under the most optimistic assumptions, the concentration of organic molecules on the early Earth would have been far too low to allow those building blocks to assemble spontaneously into full-fledged RNA molecules. RNA assembly almost certainly would have required some sort of catalytic reaction to facilitate it. Here, again, laboratory experiments offer evidence for such a process.

Experiments show that several types of inorganic minerals can facilitate the self-assembly of complex, organic molecules. Minerals of the type that geologists call *clay* may have been especially important to the origin of life.[2] Clay is extremely common on Earth and in the oceans, where it forms through simple weathering of silicate minerals; indeed, studies of the oldest mineral grains found on Earth suggest the widespread abundance of clays more than 4.4 billion years ago, so we expect clay to have been common at the time of the origin of life. Moreover, clay minerals contain layers of molecules to which other molecules, including organic molecules, can adhere. When organic molecules stick to the clay in this way, the mineral surface structure can force them into such close proximity that they react with one another to form longer chains. Laboratory experiments with this process have already produced strands of RNA up to nearly 100 bases in length. Although these strands are still quite primitive compared to what might be needed to start any sort of RNA life, the experiments suggest that clays in Earth's early oceans could have served as natural chemical "factories" that could potentially have produced self-replicating molecules of RNA. Indeed, given the millions of years and the countless grains of clay that could have facilitated chemical reactions, it seems reasonable to *expect* that self-replicating RNA molecules would at some point have been produced.

Adding further to the mix, the same clay minerals also enable the natural formation of tiny, spherical enclosures made from lipids—in essence, "pre-cells" that would have served to further concentrate potential chemical reactants. Some of the RNA strands would have ended up enclosed with these pre-cells (experiments by Jack Szostak at Harvard have demonstrated this process), thereby creating conditions for a molecular analog to natural selection: The RNA molecules that replicated faster and more accurately would have spilled out of their pre-cells and become enclosed in new ones, thereby allowing them to rapidly dominate the population. Soon, there would have been trillions upon trillions of tiny, not-quite-alive pre-cells, each containing its own self-replicating RNA molecules.

Like DNA replication, the RNA replication would have been subject to copying errors, or mutations; in fact, because modern organisms have numerous repair mechanisms that would not have existed for early life, there should have been many more mutations. The mutations would have essentially tested all sorts of variations on the replication process, inevitably leading the RNA molecules to gain complexity and evolve more efficient replication pathways. At some point, the RNA pre-cells would have become

[2] In this context, clay refers to silicate minerals with a particular physical structure; this mineralogical definition is somewhat different from what you may think of as clay in the context of pottery or sculpture.

1. Organic precursor molecules appear.

2. RNA molecules become self-replicating.

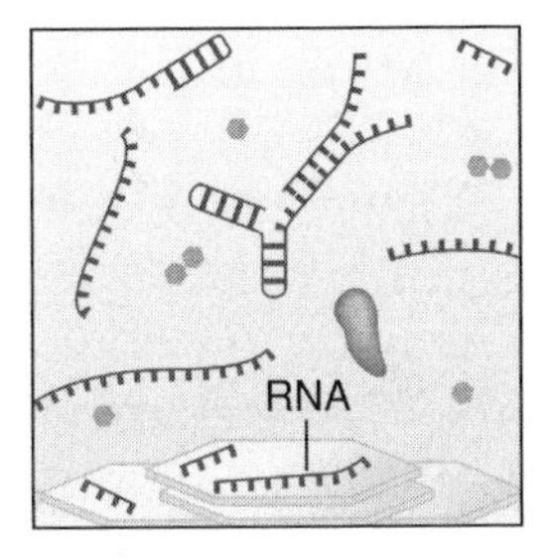

3. Membrane-enclosed pre-cells arise.

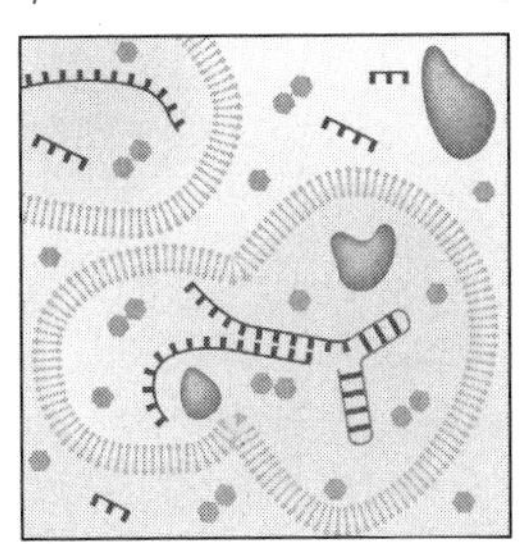

4. True cells with RNA genome appear.

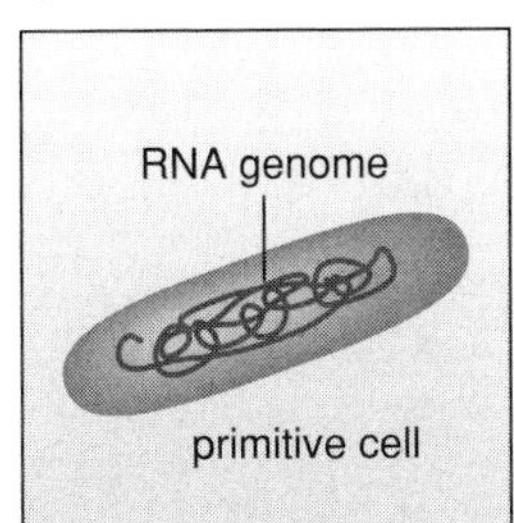

5. Modern cells with DNA genome evolve.

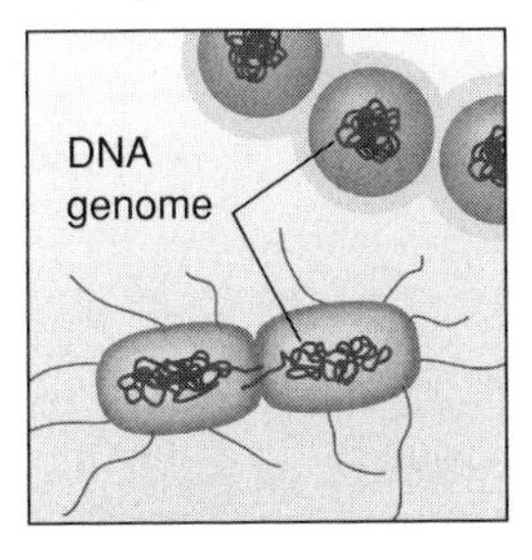

Figure 5.2. A summary of the steps by which chemistry on the early Earth might have led to the origin of life. (Illustration courtesy of Addison Wesley, an imprint of Pearson Education)

sufficiently good at reproducing and evolving that we would have considered them to be "alive." At this point, biological natural selection could take over, and it seems easy to understand why an RNA world would have eventually given way to the present DNA world: DNA is a more flexible hereditary material and is less prone to copying errors than RNA. Thus, because DNA is structurally so similar to RNA, it seems natural that the RNA molecules serving hereditary functions would have eventually evolved DNA, while other RNA molecules would have continued the cellular roles that they still play today.

Figure 5.2 summarizes the steps to life that we have outlined. We may never know for certain whether life actually originated in this way, in some similar way, or in some completely different way. Nevertheless, this scenario seems quite reasonable and perhaps even "easy" given geological time scales. It seems especially reasonable given that a number of different components of the scenario have been demonstrated in laboratory experiments. Even if life did not originate in this way, it seems that it could have—which suggests that the actual path to life must have been equally easy, or else it would have followed the path we've described. As Nobel-prize winning chemist Harold Urey once said, "If God didn't do it this way, He missed a good bet."

Regardless of God's chosen mechanism, or even of whether God really did choose, the scenario we've discussed gives us reason to think that life's early origin on Earth wasn't due to luck but instead was to be expected. In that case, we might expect the same to have happened on many other worlds, which is the major reason why so many scientists suspect that life will prove to be common in the universe.

MIGRATION SCENARIOS

Before we move on to the next topic, it's worth briefly noting that some people have proposed that life did not originate on Earth at all, but instead migrated here from some other planet. This idea, sometimes called *panspermia,* once seemed outlandish. After all, it's hard to imagine a more forbidding environment than that of space, where there's no air, no water, and constant bombardment by dangerous radiation from the Sun and stars. However, the presence of organic molecules in meteorites and comets tells us that the building blocks of life can survive in the space environment, and we've already discussed some Earth microbes (such as *D. radiodurans* and *anthrax*) that are capable of surviving at least moderate periods of time in space. It therefore seems possible that life could migrate from one planet to another, if it could hitch a suitable ride. And rides are indeed available.

We know that meteorites can and do travel from one world to another. Among the more than 20,000 meteorites that scientists have identified and cataloged, careful chemical analysis has so far revealed about three dozen with compositions that clearly suggest that they came from Mars; even more have been found that come from the Moon. Apparently, these meteorites were blasted from their home worlds by large impacts, then followed orbital trajectories that eventually caused them to land on Earth. Observations of these meteorites, along with theoretical calculations based on the amount of material blasted into space by impacts, suggests that over time the inner planets have exchanged many tons of rock. As my astronomy textbook co-author Nick Schneider puts it, Earth, Venus, and Mars have in effect been "sneezing" on each other for billions of years.

The chance of surviving the trip between planets probably depends on how long the meteorite spends in space. Once a rock is launched into space, it orbits the Sun until its orbit carries it directly into the path of another planet. Most meteorites will orbit for many millions of years before reaching Earth, even if they come from a world as nearby as Venus or Mars, and it seems highly unlikely that living organisms could survive in space for such long periods of time. However, a few meteorites are likely to be launched into orbits that cause them to crash to Earth during one of their first few trips around the Sun. For example, calculations suggest that about 1 in 10,000 meteorites may travel from Mars to Earth in a decade or less—a short enough time that some known microbial species could probably survive the journey. Note, however, that the same considerations almost certainly rule out the possibility of migration from other star systems. Under the best of circumstances, meteorites from planets around other stars would spend millions of years in space before reaching Earth, and any living organisms would almost surely be killed by exposure to cosmic rays during this time.

We therefore encounter the intriguing possibility that if Venus or Mars for some reason got life before Earth, this life might have seeded our planet, avoiding the need for an indigenous origin of life on Earth. As I'll discuss in chapter 6, we do indeed have good reason to think that the young Venus and young Mars may both have had conditions similar to those on the young Earth, making this scenario plausible, if not likely. Still, it would hardly seem to change our basic scenario for the origin of life, as it simply moves it to another world. But while migration *to* Earth might have little impact on our ideas about the origin of life, migration *from* Earth raises some interesting issues. The numerous impacts that have occurred over the past 4 billion years have surely offered many opportunities for microbes to hitch rides on meteorites blasted off Earth's surface, which means that if it were possible for Earth life to survive on any nearby worlds, we should actually *expect* to find it there. If we someday find life on Mars, for example, we may be hard-pressed to determine whether it originated there or migrated from Earth. We can only hope that the life will show enough biochemical differences or similarities with Earth life to allow us to distinguish migration from a separate origin.

GOING BEYOND MICROBES

I've made the case that we have good reason to think that microbial life will prove to be common in the universe. That would be very good for biology and probably for medicine, since comparisons between alien microbes and terrestrial microbes would surely teach us a lot. But if we want someone else to talk to, we need to find life that has gone well beyond the microbial stage. And some of the same type of evidence that suggests that getting life might be easy tells us that moving beyond microbes may be far more difficult.

Take another look at the timeline in color plate 3. From an animal-centric perspective, we might describe the history of life on Earth like this: First there were microbes, starting nearly 4 billion years ago. Then, for the first billion years or so, there were microbes. And for the next billion years after that, there were microbes. And then it almost got interesting, but not really—it's still microbes during the third billion years of life on Earth. We have to look all the way to about the last 15 percent of life's history before we find even the most primitive animals, and we can see big things like dinosaurs and mammals only during about the last 5 percent. Given that I've argued earlier that the rapid origin of life on Earth suggests that we might expect the same to happen as easily on other worlds, does the long time to animals suggest that most other worlds might never get past the microbial stage?

Maybe, but we really don't know. Although it was indeed "all microbes, all the time" for some 3 billion years or more, evolution was not sitting still. Rather, it was producing the great microbial diversity that we see among the three domains of life, Bacteria, Archaea, and Eukarya. Given all this diversity, perhaps it was inevitable that something more complex, like animals, would eventually arise. In that case, the only question is whether the 3+ billion years it took for this to happen on Earth was slow, typical, or fast. If it was slow or typical, then the same thing has probably happened on many other worlds because, as we discussed in chapter 3, there are lots of stars and planetary systems that are much older than our own Earth and solar system. But if Earth was fast, and the average time from microbes to complexity is much longer than 3 billion years, then the universe might be filled with microbial life, while more advanced life is exceedingly rare. To me, this raises the biggest unanswered question in the search for extraterrestrial intelligence. For reasons I'll discuss in chapter 9, my personal guess is that once you get animals, it's only a matter of time until you get intelligence and civilization. But I'm a lot less confident that the step from microbes to animals will prove common.

Regardless of the general answer to the question of how long it might take animal life to arise on other worlds, the rise of animals on Earth was clearly tied to the rise of oxygen in the atmosphere. Aerobic respiration with oxygen can generate cellular energy much more efficiently than anaerobic processes, so oxygen was probably necessary to the evolution of energy-intensive animals. It seems likely that photosynthesis was already occurring by 3.5 billion years ago, given that the stromatolites look so much like modern mats with photosynthetic bacteria, so it is possible that oxygen was already being released into the atmosphere by that time. However, chemical and isotopic evidence from rocks shows that oxygen did not begin to build up in the atmosphere until more than a billion years later. Apparently, chemical processes were removing the oxygen as fast as the bacteria could make it. Even once the buildup began, the limited available evidence suggests it occurred quite slowly. Although we don't know for sure, the oxygen level probably did not reach 10 percent of the overall atmospheric composition (versus its current 21 percent) until around or shortly before the time of the Cambrian explosion, when animal diversity exploded onto the scene between about 540 and 500 million years ago.[3] Perhaps the rise of oxygen helped trigger this fairly sudden evolutionary change.

Incidentally, the first clear evidence of oxygen reaching present-day levels does not appear in the fossil record until about 200 million years ago. That is when we first find charcoal in the fossil record, implying that enough oxygen was present in the atmosphere for fires to burn. In that case, if you could play Russian roulette with a time machine dial and randomly spin it to take you back to any point in Earth's history, you'd have only about a 1 in 20 chance of appearing at a time recent enough so that you could step out and breathe the air. There's an important lesson here: Our planet may be a great home to us today, but it has not always been so, and unless we take good care of it, we have no guarantees that it will remain so in the future.

RANDOM ACTS OF EVOLUTIONARY KINDNESS

Although my personal opinion is that once animal life gets going, intelligence will ultimately follow, I have to admit that it's difficult to back my

[3]It's worth remembering that this is an "explosion" only in a geological sense. The Cambrian explosion unfolded over about 40 million years, so if you went to visit Earth at any particular time during that period, you would *not* have noticed anything unusual going on.

opinion with evidence. In fact, if you look at the history of life on Earth over the past few hundred million years, you might conclude that we are here only as a result of a series of evolutionary accidents. In particular, numerous deep ice ages have exerted evolutionary pressure that probably played a big role in paving the way for our existence, as did some even more dramatic events—impacts of asteroids or comets from space. Of course, in keeping with my personal guesses, I prefer to think of these events not as accidents but rather as evolutionary acts of kindness.

Between ice ages, impacts, major volcanic eruptions, and perhaps some other events, there have indeed been a lot of random acts of kindness that have made our existence possible today. But the best known and arguably most important of these events concerns the death of the dinosaurs, and it bears a bit of discussion because it also holds implications to the possibilities of finding other civilizations.

In 1978, while analyzing geological samples collected in Italy, a scientific team led by Luis and Walter Alvarez (father and son) made a startling discovery in the thin layer of sediments that marks the Cretaceous-Tertiary boundary, or the *K-T boundary* for short (the K comes from the German word for Cretaceous, *Kreide*). The K-T boundary layer separates the sediments of the Cretaceous period, which ended 65 million years ago, from those of the Tertiary period. The boundary was already notable because it marked the extinction of the dinosaurs: Dinosaur fossils are present in the Cretaceous rocks, but not in the Tertiary. The Alvarez team discovered that the boundary layer (typically about an inch thick) is unusually rich in iridium—an element that is rare on Earth's surface but more common in meteorites. Subsequent studies found the same iridium-rich sediment marking the K-T boundary at many other sites around the world. The Alvarez team proposed a stunning hypothesis: The extinction of the dinosaurs was caused by the impact of an asteroid or comet. They calculated that it would take an asteroid about 10–15 kilometers in diameter to produce as much iridium as is apparently distributed worldwide in the K-T boundary layer.

In fact, the death of the dinosaurs was only a small part of the biological devastation that seems to have occurred 65 million years ago. The fossil record suggests that up to 99 percent of all individual living plants and animals died around that time, and this loss drove up to 75 percent of all existing plant and animal *species* to extinction. This makes the event a clear example of a *mass extinction*—the rapid extinction of a large percentage of all living species. Could it really have been caused by a random impact?

There's still some scientific controversy about whether the impact was the sole cause of the mass extinction or just one of many causes, but there's little doubt that a major impact coincided with the death of the dinosaurs. Key evidence comes from further analysis of the K-T sediment layer. Besides being unusually rich in iridium, this layer contains four other unusual features: (1) high abundances of several other metals, including osmium, gold, and platinum; (2) grains of "shocked quartz," quartz crystals with a distinctive structure that indicates they experienced the high-temperature and high-pressure conditions of an impact; (3) spherical rock "droplets" of a type known to form when drops of molten rock cool and solidify in the air; and (4) soot that appears to have been produced by widespread forest fires.

All these features point to an impact. The metal abundances look much like what we commonly find in meteorites rather than what we find elsewhere on Earth's surface. Shocked quartz is also found at other known impact sites, such as Meteor Crater in Arizona. The rock "droplets" presumably were made from molten rock splashed into the air by the force and heat of the impact. Some debris would have been blasted so high that it rose above the atmosphere, spreading worldwide before falling back to Earth. On their downward plunge, friction would have heated the debris particles until they became a hot, glowing rain of rock. The soot probably came from vast forest fires ignited by radiation from this impact debris.

The "smoking gun" for the impact is a large crater that appears to match the age of the sediment layer. The crater, about 200 kilometers across, is located on the coast of Mexico's Yucatán Peninsula, about half on land and half underwater. (It is not visible to the eye, but shows up clearly in measurements of small, local variations in the strength of gravity.) Its size indicates that it was created by the impact of an asteroid or a comet measuring about 10 kilometers across (craters are typically 10 to 20 times as large as the objects that make them), large enough to account for the iridium and other metals. It is named the *Chicxulub crater,* after a nearby fishing village.

If the impact was indeed the cause of the mass extinction, here's how it probably happened: On that fateful day some 65 million years ago, the asteroid or comet slammed into Mexico with the force of a hundred million hydrogen bombs. It apparently hit at an angle, sending a shower of red-hot debris across the continent of North America. A huge tsunami sloshed more than 1,000 kilometers inland. Much of North American life may have been wiped out almost immediately. Not long after, the hot debris raining around the rest of the world ignited fires that killed many other living organisms.

Dust and smoke remained in the atmosphere for weeks or months, blocking sunlight and causing temperatures to fall as if Earth were experiencing a global and extremely harsh winter. The reduced sunlight would have stopped photosynthesis for up to a year, killing large numbers of species throughout the food chain. This period of cold may have been followed by a period of unusual warmth: Some evidence suggests that the impact site was rich in carbonate rocks, so the impact may have released large amounts of carbon dioxide into the atmosphere. The added carbon dioxide would have strengthened the greenhouse effect, so that the months of global winter immediately after the impact might have been followed by decades or longer of global summer. The impact probably also caused chemical reactions in the atmosphere that produced large quantities of harmful compounds, such as nitrous oxides. These compounds dissolved in the oceans, where they probably were responsible for killing vast numbers of marine organisms. Acid rain may have been another by-product, killing vegetation and acidifying lakes around the world.

Perhaps the most astonishing fact is not that up to 75 percent of all plant and animal species died but that some 25 percent survived. Among the survivors were a few small mammals. These mammals may have survived in part because they lived in underground burrows and managed to store enough food to outlast the global winter that immediately followed the impact.

The evolutionary impact of the extinctions was profound. For 180 million years, dinosaurs had diversified into a great many species large and small, while most mammals (which had arisen at almost the same time as the dinosaurs) had mostly remained small and rodent-like. With the dinosaurs gone, mammals became the new animal kings of the planet. Over the next 65 million years, the small mammals rapidly evolved into an assortment of much larger mammals—ultimately including us. Had it not been for the K-T impact, dinosaurs might still rule the Earth.

The K-T layer is just one of several rock layers that reveal evidence of mass extinctions that fundamentally changed the evolutionary history of life on Earth. Some of the others may also have been caused by impacts, though the evidence is less clear; alternatively, they may have been caused by massive volcanic eruptions, or perhaps even by vast increases in mutation rates caused by external events, such as a nearby supernova that might have produced radiation that directly affected life or that might have destroyed much of Earth's protective ozone layer.

Whatever their causes, without the past mass extinctions we probably would not be here today. These random acts therefore seem kind only if they ultimately lead to beneficiaries—like us—who can appreciate the way they made our existence possible. The dinosaurs, if they had known what hit them, presumably would have been less thrilled. And this idea leads to an important question: If not for the particular series of accidents that occurred on Earth, would intelligent life ever have arisen here? If the dinosaurs were still here, for example, would they have ultimately evolved intelligence and started to build spaceships, or would Earth still be overrun with lumbering giants? Again, we really don't know, but keep in mind that impacts and other major events that can cause mass extinctions are inevitable. The dinosaurs may have been "got" by the K-T impact, but if it hadn't been that, something else probably would have got them eventually. Over hundreds of millions and billions of years, change will always occur, and my guess remains that if we hadn't come along, someone else would eventually have come in our place.

The topic of mass extinctions also holds a cautionary lesson for us. Human activity is driving numerous species toward extinction. The best-known cases involve relatively large and wide-ranging animals, such as the passenger pigeon (extinct since the early 1900s) and the Siberian tiger (nearing extinction). But most of the estimated 10 million or more plant and animal species on our planet live in localized habitats, and most of these species have not even been cataloged. The destruction of just a few square kilometers of forest may mean the extinction of species that live only in that area. According to some estimates, human activity is driving species to extinction so rapidly that half of today's species could be gone within a few centuries or less. On the scale of geological time, the disappearance of half the world's species in just a few hundred years would certainly qualify as another of the Earth's mass extinctions, potentially changing the global environment in ways that we are unable to predict. During past mass extinctions, the dominant animal species—those at the top of the food chain—have never made it through to the other side. Today, we are the dominant animal species. Perhaps our intelligence would enable us to find a way to survive while other species perish around us, but I wouldn't count on it. Those who ignore history are doomed to repeat it, and geological history tells us that perpetrating a mass extinction is not in our best interest. Unless we want to be replaced soon by the next dominant animal species—some type of insect, perhaps—then we'd be wise to heed the lesson of the past, and start doing a much better job of preserving the remarkable biodiversity upon which our survival depends.

BEYOND UFOS

If we step back to summarize what our study of life on Earth tells us about the possibilities for life elsewhere, I believe we can draw two key lessons. First, while the single example of Earth can never prove anything, since we are dealing with statistics of one, our understanding of life on this planet gives us good reason to think that *microbial* life will prove to be common throughout the universe. A century from now, I suspect we'll know of many other worlds with microbial life, including a few right here in our own solar system and many more beyond. The second lesson is that the transition from microbial to complex is much more difficult, and we do not yet understand it well enough to be able to make clear statements about whether such a transition would be rare or common. But there is one thing we can say, and it will bring us to the topic of the next chapter.

Even if you accept my guess that intelligence is an inevitable outcome once you get animal life, there's no guarantee of ever getting even that far. Moreover, life on most worlds probably *cannot* get that far, because the nature of the worlds themselves won't allow it. For example, if Venus or Mars ever had life, the current conditions on those planets would limit any surviving life to living in only a few places, such as in water-infiltrated underground rocks on Mars, or in droplets of acidic water in the clouds of Venus. It's hard to imagine anything beyond microbial life surviving under such constraints.

If life is going to get beyond the basics, it needs a planet not only where hospitable conditions make possible an origin of life, but where those conditions remain stable for the billions of years necessary to give life a chance to take itself to the next level. In our solar system, Earth is the *only* world that has had such conditions over the long haul. Indeed, because we have not yet discovered any Earth-size planets around other stars, let alone Earth-*like* planets, for the moment Earth remains the only planet we know of in the universe on which the planetary conditions gave evolution the opening it needed to produce intelligence. It is therefore time for us to explore just what it is that makes Earth so unique among the known worlds.

6

THE MAKINGS OF A TRULY GREAT PLANET

A Rock, A River, A Tree
Hosts to species long since departed
Marked the mastodon,
The dinosaur, who left dry tokens
Of their sojourn here
On our planet floor,
Any broad alarm of their hastening doom
Is lost in the gloom of dust and ages.
—Maya Angelou, from "On the Pulse of the Morning"

You've probably heard this one: The reason our planet is so great for life is the extreme good fortune of our location in the solar system. If Earth moved just a mile closer to the Sun, we would all burn up, and if it moved just a mile farther away, the oceans would freeze. I've heard this claim from so many people—students, school teachers, friends, and even preachers—that it's apparently attained the status of an urban legend.

It sounds pretty good, and like most urban legends it contains a kernel of truth: There must indeed be some distance from the Sun that would be too hot for life to survive on Earth, and some distance at which it would be too cold. But the distance isn't a mile, nor even a few million miles, as you can realize just by thinking about Earth's orbit around the Sun: Earth's orbit is not a perfect circle, but rather is an ellipse (oval) in which our distance from the Sun varies from a minimum of about 91 million miles each January to a maximum of about 94 million miles each July. Thus, according to the urban legend, our whole planet would burn up each January and freeze each July. In reality, this 3-million-mile variation in distance has virtually no effect on the weather at all, a fact that becomes obvious when you remember that the northern hemisphere has summer during the time that Earth is farthest from the Sun and winter when Earth is closest to the Sun.

In fact, based on our understanding of how the Sun produces energy, the "acceptable" range of distances for a planet like Earth must be considerably wider than this 3-million-mile range. The Sun generates the energy that makes it shine by fusing hydrogen into helium deep in its core. Each fusion reaction converts four hydrogen nuclei into one helium nucleus. The helium nucleus weighs slightly less (by about 0.7 percent) than the four hydrogen nuclei that make it, which means a small amount of mass "disappears" with each reaction. This disappearing mass turns into energy in accord with Einstein's famous formula $E = mc^2$, and this energy is what makes the Sun shine. The overall numbers are remarkable: Deep in the core of the Sun, some 600 million tons of hydrogen fuses into 596 million tons of helium *every second*, with the remaining 4 million tons being converted into energy.

You might worry that the Sun would soon run out of fuel at this rate, but the Sun is so massive that, at birth, it had enough core hydrogen to last some 10 billion years. Since the Sun is only about 4 ½ billion years old at present, it's got a long life still ahead of it. However, over billions of years the Sun does not stay perfectly steady in brightness; instead, it must gradually brighten, for a reason you can understand if you remember something from high school called the "ideal gas law." The core of the Sun is essentially an extremely hot gas, and the ideal gas law tells us that the pressure in a gas depends on the total number of independent particles flying around within it. Because every fusion reaction turns four independent particles (the four hydrogen nuclei) into just one particle (the helium nucleus), the total number of particles in the Sun is gradually decreasing with time. This gradual decrease in the number of particles causes the solar core to shrink, because there are fewer particles to generate the pressure that supports the core against the weight of overlying layers of the Sun. The slow shrinkage, in turn, gradually increases the core temperature and the fusion rate, causing the Sun to brighten gradually with time.

More detailed calculations of this process show that the Sun must be about 30 percent brighter today than it was some 4 billion years ago when life first arose on Earth. In other words, even a 30 percent increase in the Sun's brightness has not had a detrimental effect on the ability of our planet to be a home to life. This fact clearly tells us that our planet could remain much as it is even if it were moved significantly outward from its current location, since such movement would just put it at a place where the current energy of sunlight is the same as it was at Earth's distance a few billion years ago.

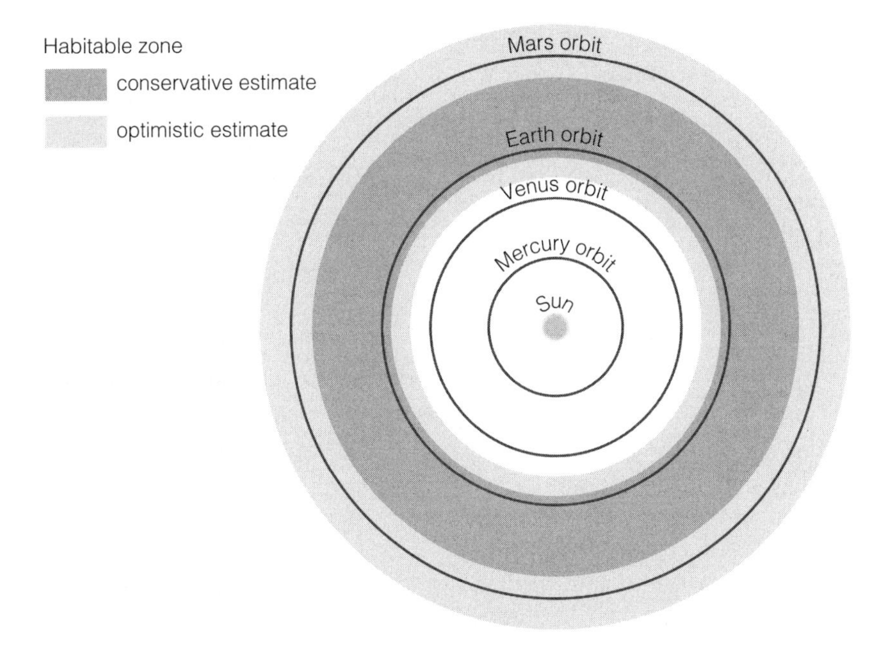

Figure 6.1. Boundaries of the Sun's present-day habitable zone—the region in which Earth could maintain surface oceans of liquid water. The narrower set of boundaries (darker shading) represents a model based on the more conservative assumptions, while the wider set (lighter shading) represents the most optimistic scenarios. (Illustration courtesy of Addison Wesley, an imprint of Pearson Education)

The region around the Sun within which Earth (or a similar planet) could remain Earth-*like*, by which we mean having conditions that allow abundant surface water, is often called the *habitable zone*. It's a bit of a misnomer, since it might well be possible for worlds outside this zone to be habitable in the sense of having life underground, or even in clouds. But it's still a useful idea, because according to what we've learned about the evolution of life on Earth, having liquid water oceans on the surface seems a clear requirement for getting past the microbial stage and on to animals and intelligence. In other words, while microbial life could be widespread on many types of worlds, only worlds located within their stars' habitable zones would seem to have a reasonable chance of allowing the long-term evolution of a huge diversity of life, perhaps ultimately leading to civilizations.

We do not yet know the precise boundaries of the Sun's habitable zone, but it certainly ranges over many millions of miles. Figure 6.1 shows the boundaries of the habitable zone under two sets of assumptions, one fairly conservative and one more optimistic. Even with the conservative assumptions, the habitable zone extends over a range of more than 40 million miles; thus, the fact that at least one planet (Earth) was born at an acceptable distance from the Sun does not seem all that remarkable. With the more optimistic assumptions, we might wonder why Mars is not a much nicer planet today, since under those assumptions it is within the present habitable zone.

The answer must be that one or more things besides distance from the Sun must be important to making Earth such a truly great planet for life. But you already knew that: If you think back to the model of the solar system that I described in chapter 3, you'll realize that for all practical purposes the Moon is at exactly the same distance from the Sun as Earth. But the Moon is dead as a doornail. To understand why, and to really understand what makes Earth unique in our solar system, we need to learn a little about why Earth is so nice and about how our neighbors went bad.

WHY EARTH ISN'T FROZEN

It's easy to measure how much energy Earth receives in sunlight; after all, it's this type of measurement that goes into calculations of how many solar panels you'd need on your rooftop in order to quit paying the electric company. Once you know how much energy Earth receives from the Sun, it's almost equally easy to calculate the surface temperature you would expect Earth to have if sunlight were the only factor. I won't take you through the calculation here (though it's easy enough that we do it in my introductory astronomy text for nonscience majors), but I'll tell you the result: Our planet should be frozen solid, right down to the bottom of equatorial oceans. Speaking more precisely, when we calculate Earth's expected global average temperature based solely on its distance from the Sun and the amount of incoming sunlight absorbed by its surface, we find that it would be about 3°F, well below the 32°F freezing point of water.

This might sound like another one of those urban myths, but it's not. It's really true: Without some sort of "blanket" to keep it warm, Earth would be too cold for life (at least on the surface). So what kind of blanket warms our planet enough for our lives to be possible? A naturally occurring type of blanket called the *greenhouse effect*.

You've probably heard of the greenhouse effect, because it's often in the news in the context of the environmental issue known as global warming. However, the greenhouse effect itself is not a bad thing, since it is what makes our planet livable. It's easy to understand how it works.

Sunlight consists mostly of visible light, which passes easily through most atmospheric gases and reaches Earth's surface. Some of this visible light gets absorbed by the ground, while the rest is reflected back to space. The ground must return the energy it absorbs back to space, because if it didn't the energy would make the ground heat up very rapidly. However, the fact that the ground doesn't glow in the dark tells us that the ground does not return the energy in the same visible light form that it absorbs it. Instead, the ground returns the energy in the form of infrared light—an invisible form of light that has wavelengths somewhat longer than those of visible light.

The greenhouse effect works by "trapping" some of the infrared light, thereby slowing its return to space. This trapping occurs because some atmospheric gases can absorb the infrared light. Gases that are particularly good at absorbing infrared light are called *greenhouse gases.* The most important greenhouse gases are water vapor (H_2O), carbon dioxide (CO_2), and methane (CH_4). These gases absorb infrared light effectively because their molecular structures make them prone to begin rotating or vibrating when struck by an infrared photon (an individual "piece" of light); diatomic molecules, such as nitrogen (N_2) and oxygen (O_2), generally cannot rotate or vibrate in these ways and hence do not absorb infrared light.

After a greenhouse gas molecule absorbs the energy of an infrared photon, it quickly releases the energy by emitting a new infrared photon. However, the new photon will be emitted in some random direction that is unlikely to be the same direction from which the original photon came. The new photon is then usually absorbed by another greenhouse gas molecule, which does the same thing. The net result is that greenhouse gases tend to slow the escape of infrared radiation from the lower atmosphere, while their molecular motions heat the surrounding air. In this way, the greenhouse effect really does work like a blanket. You stay warmer under a blanket not because the blanket itself provides any heat, but rather because it slows the escape of your body heat into the cold outside air. In the same way, the greenhouse effect keeps Earth's surface (and lower atmosphere) warmer than they would be otherwise because it slows the escape of the heat radiated by the ground back toward space. Figure 6.2 summarizes the process.

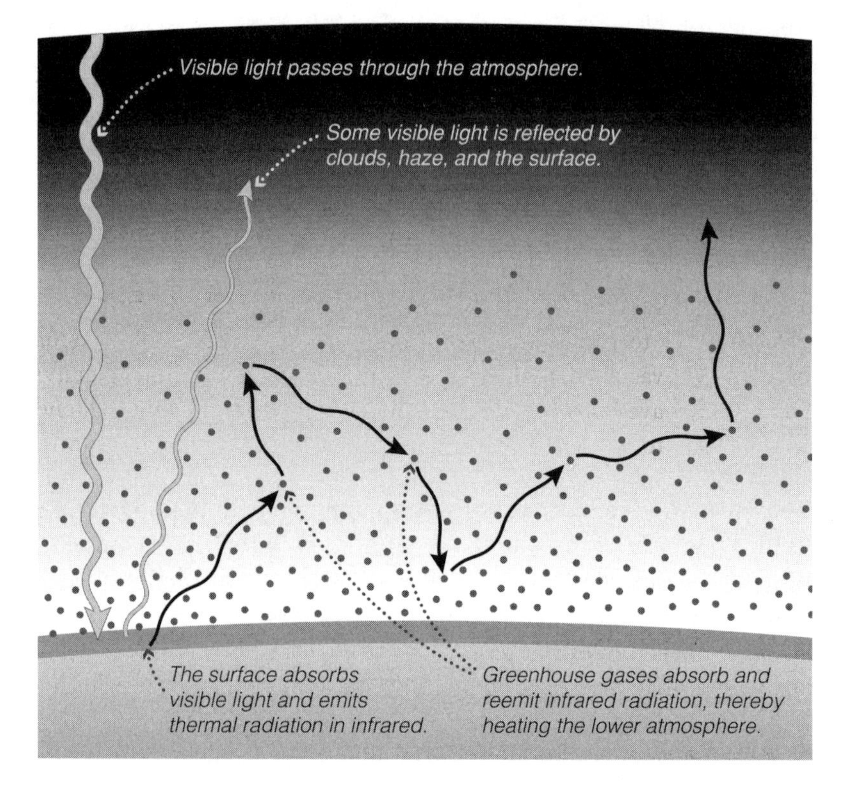

Figure 6.2. The greenhouse effect makes the surface (and lower atmosphere) much warmer than it would be without greenhouse gases such as water vapor, carbon dioxide, and methane. Without the naturally occurring greenhouse effect, Earth would be too cold to have liquid oceans. (Illustration courtesy of Addison Wesley, an imprint of Pearson Education)

We therefore have a "simple" answer to the question of why Earth is warm enough for life: Our planet's atmosphere contains just the right amount of greenhouse gases to make the temperature warm enough for liquid oceans, but not so warm that the oceans would start to boil away. But it doesn't take long to realize that this simple explanation really creates more questions than it answers. For example, given that all the greenhouse gases are essentially trace components of an atmosphere made almost entirely of nitrogen (77 percent) and oxygen (21 percent), how is it that the greenhouse gas content is so finely tuned to the needs of life? And given that the Sun has brightened by some 30 percent over the past 4 billion years, how is it that our planet has stayed within a temperature range acceptable to life throughout that time? Remarkably, the answers to these questions come down to the fact that Earth has a self-regulating, natural thermostat that automatically adjusts the atmospheric concentration of carbon dioxide as needed to keep temperatures within a reasonable range.

It sounds almost miraculous, and I won't argue with you if you are inclined to believe it is. But regardless of whether it is evidence of God's handiwork, we now understand how the miracle occurs through a combination of processes that are all quite natural. The easiest way to understand the basic idea is to start with our planet as it was a long, long time ago, shortly after it was born.

Earth must have been born without any atmosphere to speak of, because its gravity was too weak to hold on to the hydrogen and helium gas that surrounded the planets as they formed. However, Earth must have had gases trapped within its interior, brought to our planet by planetesimals that contained ices (which are essentially frozen gases). These gases were held under pressure in the interior in much the same way that the gas in a carbonated beverage is trapped in a pressurized bottle. When molten rock erupts onto the surface as lava, the release of pressure violently expels the trapped gas in a process we call *outgassing*. Outgassing released the gas that made up Earth's early atmosphere. Some volcanoes today are apparently little different from those of long ago, so by studying them we learn the composition of the gas that was expelled to make the early atmosphere. The major gases released by such volcanoes are water vapor (H_2O), carbon dioxide (CO_2), nitrogen (N_2), sulfur-bearing gases (H_2S or SO_2), and hydrogen (H_2).[1]

Notice that the first two gases on the list, water vapor and carbon dioxide, are both greenhouse gases. These warmed our planet enough to raise the global average temperature above the freezing point, even though the Sun was dimmer at the time. And this led to an amazing chain of events. Because the temperature was warm enough for liquid water, the water vapor in the atmosphere condensed to make rain, and the rain filled Earth's oceans. Carbon dioxide also dissolved in rain water, and as the oceans formed it became dissolved within them—even today, there is about 60

[1] The outgassed hydrogen eventually escaped to space (though the precise rate at which it escaped is a matter of debate at this time), and the fate of the water and carbon dioxide is described below. That primarily leaves nitrogen, which is why Earth's atmosphere is made mostly of nitrogen. As described in the prior chapter, the oxygen in Earth's atmosphere was released by life, through photosynthesis.

times as much carbon dioxide dissolved in the oceans as in the atmosphere. Once in the oceans, the carbon dioxide underwent chemical reactions with dissolved minerals that made *carbonate* rocks, such as limestone. The vast amount of water in Earth's oceans therefore tells us how much water vapor was released by volcanism, and the amount of carbon dioxide now found in carbonate rocks tells us how much carbon dioxide was released. Hold onto your hats, because the answer is shocking: Carbonate rocks contain close to 200,000 times (more precise estimates put it at about 170,000 times) as much carbon dioxide as Earth's atmosphere. If all this carbon dioxide had somehow remained in the atmosphere instead of having been dissolved in the ocean and trapped in rock, Earth's surface would be baked to a crisp, and certainly would be lifeless.

You may wonder: If this much carbon dioxide got trapped in the carbonate rock, why didn't *all* of it get trapped, and why is the amount remaining in our atmosphere "just right" to keep the oceans liquid? The answer has to do with the way in which carbonate rock is recycled. As we'll discuss momentarily, some carbonate rock is continually being melted, causing it to release its carbon dioxide back into the atmosphere. This carbon dioxide can then be absorbed into new carbonate rock, but the precise rate at which the new rock forms depends on how rapidly carbon dioxide is dissolved in the ocean. This rate turns out to be very sensitive to temperature: Higher temperature increases the rate of absorption. If our planet starts to warm up—for example, as we might expect it to have done as the Sun gradually brightened with time—the increased evaporation of ocean water means more rain, which means more atmospheric carbon dioxide dissolves in the rain and ends up in the ocean. This effectively pulls carbon dioxide out of the atmosphere, reducing the strength of the greenhouse effect and cooling the planet back down. Conversely, if the planet cools, the lower rainfall allows carbon dioxide to build up in the atmosphere, strengthening the greenhouse effect and warming the planet back up.

The last piece of this puzzle known as the *carbon dioxide cycle* is the question of why the carbonate rock eventually gets recycled rather than just piling up at the bottom of the ocean. The answer lies with the basic geology of Earth itself, and in particular with the process that we call *plate tectonics*. Earth's outer layer of fairly cool and rigid rock, technically know as its *lithosphere* (which encompasses the thin crust and the very upper part of the mantle) extends downward to a depth of only about 60 miles. There, it essentially "floats" on top of the underlying, warmer rock of Earth's mantle. Because Earth is hot inside, the mantle rock is not motionless, but instead moves gradually in a pattern that we call *convection*—the

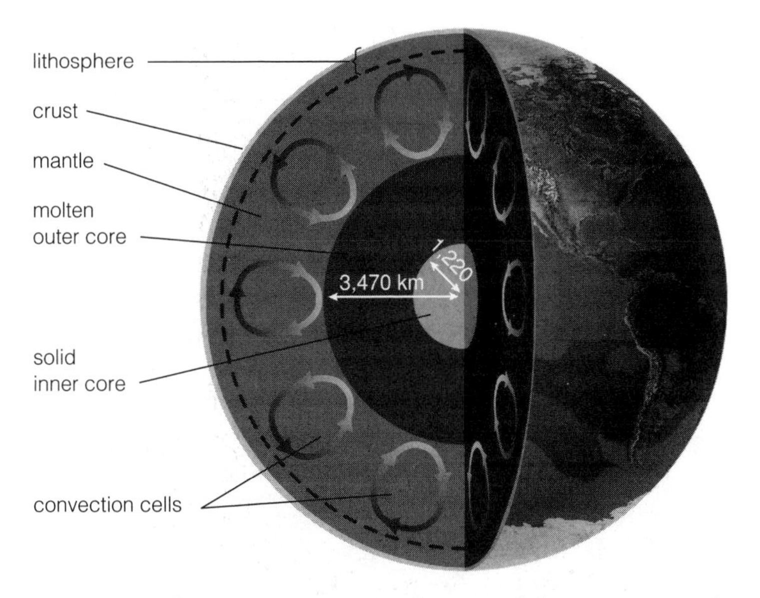

Figure 6.3. Earth's interior structure (determined from seismic studies) with the lithosphere and the underlying mantle convection indicated. (Illustration courtesy of Addison Wesley, an imprint of Pearson Education.)

same pattern that we see in weather when warm air rises and cool air falls. That is, the warmer mantle rock from deep below gradually rises up toward the base of the lithosphere, while the cooler mantle rock near the top sinks back down. Figure 6.3 shows a diagram of Earth's interior, with looped arrows representing the mantle convection. By the way, if you're surprised by the idea that the "solid" rock of the mantle can flow, don't be; solids do flow, just a lot more slowly than liquid. The popular toy Silly Putty provides a good analogy: The putty can feel pretty solid, especially when it is cold; you can even form it into a ball and bounce it. But if you put a pile of it on a table or inside its "egg" container, after a few days you'll see that it has flowed slowly outward. The mantle flows more slowly still: Typically, mantle rock flows at a rate of perhaps ten centimeters per year, slow enough that it would take about 100 million years for a particular piece of rock to be carried from the base to the top of the mantle.

Mantle convection has apparently caused Earth's lithosphere to fracture in numerous places, so that rather than being solid the lithosphere is broken into about a dozen "plates." These are the plates you hear about whenever you hear of an earthquake, since most (but not all!) earthquakes occur

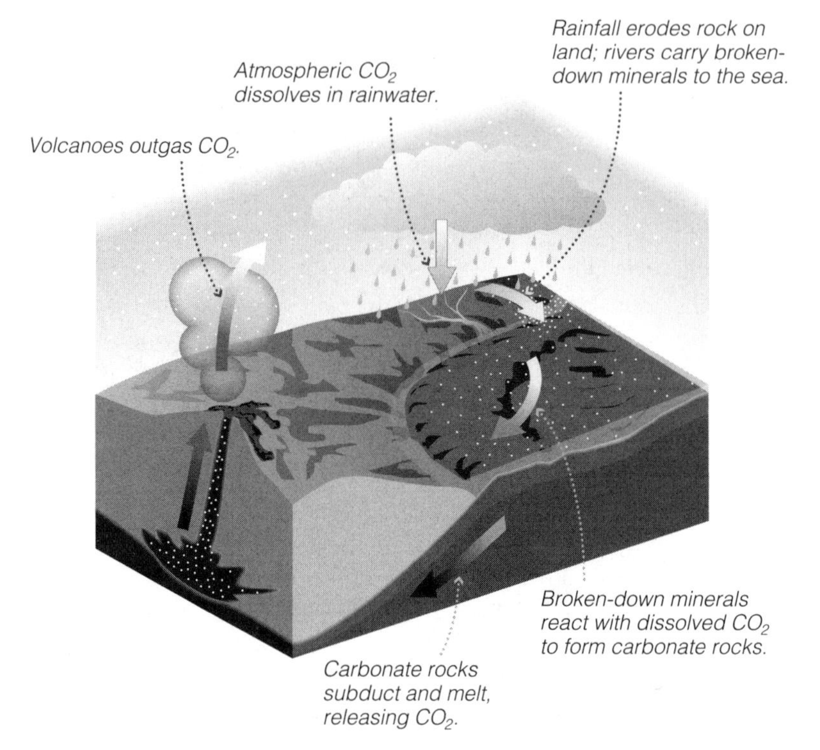

Figure 6.4. This diagram shows how the carbon dioxide cycle continually moves carbon dioxide from the atmosphere to the ocean to rock and back to the atmosphere. Note that plate tectonics (subduction in particular) plays a crucial role in the cycle. The cycle acts as a natural thermostat for our planet.

along plate boundaries. At some plate boundaries—in particular, those that occur along the mid-ocean ridges—new mantle material comes upward to the surface, creating new seafloor. Over millions of years, this new seafloor moves across the ocean bottom until it reaches the boundary between a seafloor plate and a continental plate. The seafloor rock then plunges back down into the mantle (a process called *subduction*), where it can melt and, most important for our discussion, release the carbon dioxide that had become trapped in seafloor rock.

So there you have it, and if I've lost you in the details I'll just summarize the bottom line: Thanks to plate tectonics and the existence of our oceans, the amount of carbon dioxide in Earth's atmosphere—and hence the strength of the greenhouse effect—is automatically regulated by the carbon dioxide cycle. Figure 6.4 summarizes how it works.

Now, I know I shouldn't play pop psychologist, but I have to ask: How does it make you feel to know that our planet can regulate its own temperature so nicely? On the one hand, it should make you feel very good; after all, without this natural regulation we could not be here today. But don't let it make you feel too good, because while Earth can indeed take care of itself, it does so on *geological* time scales, not human time scales. I'm not trying to be facetious; there are some people out there who might hear this stuff and say, "See, we have nothing to worry about with global warming, because Earth's climate will automatically fix whatever damage we cause." Don't be fooled. On time scales of decades, centuries, millennia, and even longer, other factors can easily overwhelm the natural thermostat, which is why Earth has suffered through so many ice ages and warm periods in the past. If we choose to muck with our planet enough to wreck our civilization, nature ain't going to step up to save us.

I'll have more to say about global warming and its relevance to the search for life in the universe a little later in this chapter. First, however, we need to turn our attention to neighboring worlds. For almost everything that has gone "right" with Earth, our neighbors offer examples of what can instead go wrong. They therefore hold critical lessons to our understanding of the prospects for finding Earth-like planets elsewhere in the universe, as well as to the future of our own species.

LESSONS FROM OUR NEIGHBORS I: THE DEAD MOON

Although the Moon is clearly within the Sun's habitable zone, it is *not* habitable, because it has no liquid water anywhere. In fact, the Moon lacks water of any type, except perhaps for some ice that remains permanently frozen in craters near the Moon's poles.[2] The Moon's almost complete lack of water is thought to be a result of the way it was formed. Recall that the leading hypothesis holds that the Moon formed from material blasted out of Earth's outer layers by a giant impact; the heat of the impact would have vaporized all the water from these layers, and once it was in gaseous form

[2] Comet impacts must bring water ice to the Moon, so the polar craters that lie in perpetual shadow might still have this ice at their bottoms. Elsewhere on the Moon, sunlight causes the ice to sublimate away. If it really exists (the current evidence is unclear), this polar ice could prove useful to future human colonists, who could use it as a local source of water.

the water would have been unable to participate in the process of accretion that assembled the Moon from the impact debris. But it wouldn't have mattered even if the Moon had been born with water, because the Moon has a much more severe problem when it comes to habitability: It is too small.

Now, size isn't everything when it comes to life. As we'll discuss in chapter 7, some relatively small moons of the outer solar system could potentially be homes to life. But these worlds owe their good fortune to two things that the Moon lacks: a composition that includes vast amounts of ice and a heating source (called tidal heating) that can be significant only when a planet has multiple moons. For an isolated, rocky world like the Moon, size alone explains its lack of habitability.

Why is size so important? Just think about baked potatoes: The small ones cool more quickly than the big ones. The same basic idea holds true for planets and other worlds. All worlds start out with fairly hot interiors, primarily as a result of the heat deposited by the collisions that form them and of heat released by the decay of radioactive elements within them. Over time, this heat gradually escapes to space. Because the total amount of heat within a world is proportional to the world's total interior *volume,* and because the heat escapes only through the world's *surface,* the rate at which heat leaks away is proportional to what we call the *surface area to volume ratio.* Small objects always have greater surface area to volume ratios than bigger ones,[3] which is why they lose heat more quickly. Incidentally, the same basic idea explains why crushed ice cools a drink faster than ice cubes: The small bits of crushed ice have a greater total surface area than the same volume of ice cubes, and more surface area means more ice in contact with the liquid drink that you are trying to cool.

With a radius barely 1/4 that of Earth, the Moon has by now cooled so much that its interior can't support volcanism; in fact, the Moon's interior has probably been this way for at least 3 billion years by now. With no volcanism, there is no outgassing that could release atmospheric gases. Moreover, the Moon's small size means weak gravity, so even if the Moon had released some atmospheric gas long ago, the gases would have quickly escaped into space. The cool interior has also allowed the Moon's rigid lithosphere to thicken to the point that it now extends a large fraction of the way toward the Moon's center, ensuring that it is too strong to fracture into plates. And even if it were fractured, the lack of internal heat means no

[3] Mathematically, volume is proportional to radius cubed while surface area is proportional to radius squared. The surface area to volume ratio is therefore proportional to 1/radius, which means it is larger for smaller objects.

mantle convection, and hence no plate tectonics. Putting all these ideas together, we see that the Moon is airless and lifeless because its small size means that it has no way to release or recycle gases, and any gas from long ago is long gone. The planet Mercury is dead for the same reason, since its radius is barely 1/3 that of Earth. Of course, from the standpoint of life, Mercury was probably doomed anyway, because it is located too close to the Sun.

The lesson of the Moon should be clear: When it comes to the making of an Earth-like world, size matters. The world has to be large enough to retain internal heat for billions of years, so that volcanism can release interior gases to create an atmosphere, and so that plate tectonics can support a gas recycling system that can serve as a planetary thermostat.

LESSONS FROM OUR NEIGHBORS II: THE FREEZE-DRYING OF MARS

Mars is intermediate in size between the Moon and Earth, with a radius slightly over half that of Earth. We therefore expect it to retain at least some internal heat, but not nearly as much as Earth. We cannot yet study Martian history by digging through layers of rock, but thanks to the many spacecraft we've sent to Mars, we can use features of the Martian surface to learn about how its geology has changed with time. Visually, Mars presents a complex story.

The complexity should become clear with your first glance at the map of Mars shown in color plate 4. Notice, for example, that the southern hemisphere is heavily cratered. Because most craters were made during the ancient heavy bombardment, the dense cratering tells us that this surface has not been repaved by volcanic eruptions or other processes over the past 4 billion years or so. However, the northern hemisphere tells a different story, as its relative absence of craters suggests geological activity well after the end of the heavy bombardment. Smoking guns for volcanic activity take the form of the many large volcanoes that dot the Martian landscape, including Olympus Mons and the other nearby volcanoes of the Tharsis Bulge.

No one knows why Mars shows such a dichotomy of landscapes between its northern and southern hemispheres, but for our purposes it may not matter. Since Mars has volcanoes, it has had outgassing, so we have an explanation for why Mars has an atmosphere. By counting craters on the slopes of the volcanoes, we can estimate how long it has been since the volcanoes have been active. The answer is that it's been quite a while—probably at least tens to hundreds of millions of years—but not long enough to conclude that the volcanoes are dead. In fact, one of the Martian meteorites that we have found here on Earth shows evidence that Mars has had volcanic activity within the past 200 million years, a short enough time compared to the age of the solar system that it is a virtual certainty that at least some of the volcanoes are still active. It may be tens of millions of years or longer between eruptions, but Mars apparently still retains some internal heat.

The real surprise comes when we examine close-up photos, both from spacecraft orbiting Mars and from the robotic landers we've sent to the surface. Orbital photos show numerous features telling us that Mars once had flowing water (figure 6.5). These features include ancient riverbeds, lake beds at the bottoms of craters, and crater rims that appear to have been eroded by rainfall. Studies of the surface conducted by the *Spirit* and *Opportunity* rovers show numerous minerals and rock outcroppings that appear to have formed in water, adding further support to the basic idea that Mars was once a wet planet.

Why should the wetness be a surprise? It's because Mars does not have any liquid water on its surface today. Although Mars has an atmosphere, the air is very thin compared to air on Earth. In fact, the air pressure on Mars is so low that if you went outside without a spacesuit, you'd die within a minute or two due to the pressure difference between your inside tissues and the outside air. The low air pressure also makes liquid water unstable on the Martian surface: If you put on a spacesuit and took a cup of water outside a pressurized spaceship on Mars, the water would almost immediately either freeze or boil away (or some combination of both). When temperatures are warm enough, water ice on Mars does what dry ice (which is frozen carbon dioxide) does on Earth: It sublimates directly from the solid phase into the gas phase, without first melting into liquid.

We therefore reach an important conclusion: Some time in the distant past—probably at least 2–3 billion years ago, based on crater counts around rivers—Mars underwent a dramatic and permanent change in climate. It changed from being a world that must have been at least somewhat Earth-like, with flowing rivers, lakes, and perhaps even an ocean, into the cold, dry planet that it is today. In other words, Mars apparently got off to a good start in terms of potential for being a world with life like that on Earth, but something then went terribly wrong.

To understand what went wrong on Mars, we need to think more deeply about how Mars was different in the past. In order to have had liquid water,

Figure 6.5. This photo, taken by NASA's *Viking Orbiter* looking down on Mars, shows what appear to be dried-up river beds. Toward the top of the image we see many individual tributaries, which merge into the larger "river" near the lower right. Counts of craters near the channels indicate that they formed at least 2–3 billion years ago.

the air pressure must have been much greater than it is today; that is, Mars must have had a much thicker atmosphere in the past than it does now. In addition, Mars must have been warmer in the past, since it is now generally too cold for liquid water, which means it must have had a much stronger greenhouse effect. We can explain both the greater pressure and the stronger greenhouse effect by assuming that Mars once had an atmosphere containing a few hundred times as much carbon dioxide as it does today.

This idea actually makes sense. Even today, the Martian atmosphere is composed mostly (about 95 percent) of carbon dioxide, so we know that Martian volcanoes must have outgassed both water vapor and carbon dioxide. If we assume that Martian volcanoes release these gases in the same proportions as Earth volcanoes, then Mars should indeed have had plenty of water vapor to condense as rain into rivers and lakes and possibly oceans, and plenty of carbon dioxide to make a much thicker and warmer atmosphere than it has today. The real question is where all this gas went.

Before I tell you the likely answer, I need to tell you of a point of scientific controversy that I've glossed over so far: Models of the Martian climate suggest that carbon dioxide alone cannot produce a strong enough greenhouse effect to fully explain the warming needed in the past, especially since the Sun was dimmer at that time. The most likely explanation for this discrepancy (at least in my opinion) is that other greenhouse gases, such as methane, made up the difference. However, some scientists suspect that the discrepancy cannot be so easily explained, and therefore argue that Mars never had an extended Earth-like period and instead had only intermittent periods during which it was warm enough for water to flow. Since life seems more likely to have started with an extended wet period, the resolution of this debate could have important implications to the possibility of life on Mars. So keep this issue in mind for when we discuss life on Mars in the next chapter, and also keep it in mind as another lesson in the idea that science is a way of getting people to come to agreement: Science will eventually resolve this debate one way or the other, because further studies of the Martian surface, perhaps along with radiometric dating of Martian rocks that we'll someday collect and study, will tell us whether the wet period was extended or intermittent.

Regardless of the details, the fact that Mars once had flowing water and no longer does tells us that it really did lose a lot of atmospheric gas; in fact, it probably lost hundreds of times as much gas as it still has today. Our best guess about how it lost this gas comes back to Mars's relatively small size, but in a way somewhat less direct than the escape to space through which the Moon or Mercury would have lost their early atmospheric gas.

Mars is large enough so that we wouldn't expect "heavy" (in terms of molecular weight) gases such as water vapor or carbon dioxide to be able to escape to space on their own. However, gases can be lost to space in other ways as well. For example, gas can be blasted into space by impacts, and Mars's relatively small size would have allowed gas to be blasted away like this more easily than gas could be blasted away on a world the size of Earth. In addition, the Sun blows a constant stream of charged particles into space, making up what we call the *solar wind,* and these particles can essentially strip gas molecules out of a planetary atmosphere as they pass by. We have some reason to think that this type of "solar wind stripping" was the primary mechanism by which Mars lost its atmosphere.

"Wait," you might say, "Earth is even closer to the Sun than Mars, so if the solar wind stripped away Mars's atmosphere, why didn't it do the same to Earth's?" Easy: Earth's atmosphere is protected from the solar wind by our planet's global magnetic field. The same magnetic field that makes your compass needle point north creates a large *magnetosphere* surrounding our planet; the magnetosphere deflects most solar wind particles past our planet, and channels those that remain toward the poles, where they create the dancing lights in the sky known as the auroras. And not to put words in your mouth, but your next question must be, "OK, then, but why doesn't Mars also have a magnetic field to protect it?" Here's where size comes in.

If you think back to experiments you probably performed in elementary school, you'll remember that there are two basic ways to get a magnetic field. First, there are objects—the iron bars we call magnets—whose atoms are arranged in such a way that they essentially have permanent magnetic fields. But while these magnets may be impressive on your refrigerator, their magnetic fields are fairly weak compared to those we can create with the second type of magnet: electromagnets, which you can make by coiling a wire around a metal screw and attaching the ends to the positive and negative terminals of a battery. Electromagnets work because charged particles going in circles generate magnetism, and they can be very powerful. Earth's magnetic field is essentially generated by an electromagnet, but instead of having the charged particles moving in circles due to a battery, they move in circles due to their flow within Earth's liquid outer core. The outer core is liquid because Earth has enough internal heat to melt metal in this region of the interior, and the freely moving electrons within this molten metal move in circles through a combination of Earth's rotation and convection within the liquid outer core. To summarize, Earth has a global magnetic field because it is hot enough to have a layer of convecting molten metal inside it and it is rotating fast enough to get that molten metal really flowing. Mars rotates at about the same rate as Earth, but it has cooled enough so that any molten metal in its core is no longer convecting enough to generate a protective magnetic field.

With that in mind, the likely history of the Martian atmosphere is as follows: Long ago, when Mars was still very hot inside, volcanism released gases that gave Mars an Earth-like atmosphere with a strong greenhouse effect, allowing water to flow on the surface. During that time, Mars had a convecting molten region in its core much like that in Earth's core today, and this region generated a magnetic field that protected the early Martian atmosphere. However, because Mars is smaller than Earth, it has lost much more of its internal heat through time. By some time between about 2 and 3 billion years ago, Mars had cooled enough so that convection stopped in its metal core, and the magnetic field went away. With the magnetic field gone, the Martian atmosphere was left vulnerable to the solar wind, which gradually stripped the atmospheric gas away. This gas loss weakened the greenhouse effect until the Martian surface essentially froze over, and the decreased pressure made it impossible for liquid water to be stable on the surface today.

This freeze-drying of Mars holds at least two important lessons as we seek to understand what makes a world like Earth so "right" for life. First, in addition to reinforcing the general lesson about the role of size that we learned from the Moon, it also tells us that a world with long-term habitability probably also needs a decent rotation rate, since both size (for internal heat) and rotation are necessary to generate a protective magnetic field. Second, it tells us that worlds can get off to promising starts in terms of prospects for life, but then undergo dramatic change. While we have no reason to think that Earth could ever freeze in the same way that Mars did, the idea that things can change dramatically should be a cautionary tale as we alter the balances that maintain the climate here on our world.

LESSONS FROM OUR NEIGHBORS III: THE OVERGREENING OF VENUS

You know how some people like to shout obscenities in foreign languages, presumably in hopes that others somehow won't notice? Well, here's something else you can try: The next time you want to tell someone where you think they should go, tell them to "go to Venus." Because if you're looking for the real location of hell, Venus is your best bet in this solar system. The surface temperature is an incredible 880°F—hotter than your self-cleaning oven and easily hot enough to melt lead—and this temperature persists planet-wide, both day and night. All the while, an extremely thick atmosphere bears down on the surface with a pressure 90 times that on Earth's surface—equivalent to the pressure at a depth of more than half a mile in the oceans. Besides this crushing pressure and searing temperature, the atmosphere of Venus contains sulfuric acid and other toxic chemicals. Venus is certainly worthy of scientific study, but you'll have a hard time finding human volunteers to set up shop there.

It might be tempting to attribute this extreme heat directly to Venus's proximity to the Sun, but a little thought shows that this idea doesn't work. For one thing, Venus's average surface temperature is hotter than Mercury's, even though Mercury is a lot closer to the Sun. Getting more precise, Venus is only about 30 percent closer to the Sun than Earth, a distance difference that makes the intensity of sunlight about twice as much (because the intensity of sunlight follows an inverse square law with distance). Twice the sunlight might sound like a lot, but you get that much change in sunlight intensity on Earth just by moving from the equator to the Arctic Circle. The equator is warm, but it's not Venus.

In fact, when you remember that Earth would be frozen if not for the greenhouse effect, you might guess that Venus's closer distance to the Sun would put it in about the right place to be a tropical planet. Indeed, past generations of science fiction writers sometimes pictured it that way, populating Venus with dinosaurs roaming through steaming forests. Venus doesn't look like this because its atmosphere is not like Earth's. Instead, its thick atmosphere is made almost entirely of carbon dioxide, and this carbon dioxide generates an extreme greenhouse effect that causes the high temperature. Given that we've already seen that the greenhouse effect is a very good thing for life on Earth, since we could not be here without it, Venus stands as a clear example of the fact that it's possible to have too much of a good thing.

To understand why Venus has such a baking hot greenhouse effect, we need to start from the idea that, at least on the inside, Venus is a lot like Earth. Venus is only about 5 percent smaller in radius than Earth, a difference in size that seems unlikely to cause any fundamental difference in planetary character or internal heat. Venus's average density is also very similar to Earth's, suggesting that it has about the same overall composition—just as we should expect for a world that formed in the same general region of the solar system. Once we accept that Venus is a lot like Earth in these ways, we can understand what happened to Venus just by asking what would happen to Earth if we moved it to Venus's orbit.

At first, things might not look so bad on our moved Earth. The greater intensity of sunlight would almost immediately raise the global average temperature from its current 59°F to about 113°F. It would be hot, but for most of the planet no worse than Las Vegas on a summer day. However, the weather would rapidly turn for the worse. The higher temperature would lead to increased evaporation of water from the oceans, and at the same time the atmosphere would increase its capacity for holding water vapor before the vapor condensed to make rain (think of how much more humid hot days are than cold). This combination would substantially increase the total amount of water vapor in Earth's atmosphere. Now, remember that water vapor, like carbon dioxide, is a greenhouse gas. The added water vapor would therefore strengthen the greenhouse effect, driving temperatures a little higher. The higher temperatures, in turn, would lead to even more ocean evaporation and more water vapor in the atmosphere—strengthening the greenhouse effect even further. In other words, we'd have a positive feedback loop in which each little bit of additional water vapor in the atmosphere would mean higher temperature and even more water vapor. The process would careen rapidly out of control, resulting in what we call a *runaway greenhouse effect.* The runaway process wouldn't stop until our planet became so hot that the oceans would be completely evaporated and the carbonate rocks would have released all their carbon dioxide back into the atmosphere. At that point, the atmosphere of our moved Earth would contain almost 200,000 times as much carbon dioxide as it does today—which is about the same amount that we find in the atmosphere of Venus. In short, moving Earth to Venus's orbit would essentially turn our planet into Venus.

So now you know how Venus went bad. It was close enough to Earth in size and composition that it should have had similar volcanic outgassing and a similar early atmosphere. Indeed, because the intensity of sunlight from the young Sun should have been roughly the same at Venus's orbit then as it is at Earth's orbit today, it's quite possible that Venus got off to a good start, with water vapor condensing to make rain and form oceans. It's even possible that Venus had plate tectonics and a carbon dioxide cycle in its early years, though it's unlikely we'll ever be able to find out, because any rocks that might tell the tale have probably undergone too much change due to the high heat over the years. Whether or not Venus had an Earth-like early history, the brightening Sun doomed it to suffer a runaway greenhouse effect.

To close the loop on Venus, we need to ask what happened to the water vapor that once was released as atmospheric gas. Inventories made by spacecraft show that Venus has virtually no water in any form today, except in some clouds at altitudes high enough that temperatures have dropped to the point where acidic droplets can condense. There's probably not even water trapped in the interior any more, because the heat would by now have baked it out of the crust and mantle and there's no way to recycle it back in. The leading hypothesis for the disappearance of the water invokes ultraviolet light from the Sun. Venus lacks an ozone layer (because it lacks life and therefore lacks oxygen), so water vapor in the atmosphere is vulnerable to ultraviolet light, which breaks water molecules apart. The hydrogen from the water molecules then escapes to space, which means the water cannot be reconstituted. As a result, the runaway greenhouse process is not reversible: The water is gone for good, so Venus could not recover even if we moved it to Earth's orbit.

The runaway greenhouse effect sealed Venus's fate, but we might ask whether Venus could have been Earth-like if it formed a little farther from the Sun—far enough so that it would still be within the habitable zone today. The answer is a definite maybe. Although Venus is a lot like Earth, two things would seem to work against its having ever become as good of a home to life as Earth. First, Venus rotates too slowly—about once every 243 days, and in a "backward" direction compared to its orbit—to generate a protective magnetic field, leaving its atmosphere vulnerable to stripping by the solar wind. Venus has probably lost even more gas to solar wind stripping than Mars (because it is closer to the Sun), but its atmosphere is so thick that this gas loss has been negligible by comparison. However, if Venus had an atmosphere as thin as Earth's—as it might have if it had never undergone a runaway greenhouse effect—the percentage gas loss could have been significant. The second potential problem is that while some type of tectonics operates on Venus—the entire planet has very few impact craters, telling us that it has somehow been repaved over time—it is clearly not plate tectonics of the same type that operates on Earth. Without plate tectonics, Venus would seem to lack the ability to have a climate-regulating thermostat.

However, before we take these two strikes against Venus and conclude that it would be out as a habitable world even if it had formed in the same place as Earth, we might ask *why* Venus rotates so slowly and lacks plate tectonics. In fact, both strikes may be *consequences* of the runaway greenhouse effect, rather than intrinsic properties that Venus was born with.

Venus's slow, backward rotation may have arisen from a drag effect created by interactions between tidal forces from the Sun and the extremely thick atmosphere. This hypothesis is relatively recent and still controversial, but it suggests that Venus would have rotated at a more "normal" rate—that is, a rate similar to the rates at which Earth and Mars rotate—if it had not undergone the runaway greenhouse effect. Remember that, if not for the runaway process, we would expect Venus to have ended up with oceans of water and with its carbon dioxide trapped in carbonate rock, and hence with an atmosphere as thin as that of Earth. With a much thinner atmosphere, any atmospheric drag effect would have been minimal, leading some scientists to speculate that Venus would then have ended up with a "normal" rotation rate. In that case, because Venus has essentially the same interior composition and internal heating as Earth, Venus would have an Earth-like magnetic field protecting this atmosphere.

Venus's lack of plate tectonics may be a consequence of the fact that the high surface temperature has baked out water from its crust and upper mantle. This drying of the rock would have strengthened and thickened Venus's lithosphere, thereby making it resistant to the fracturing that occurred on Earth; the high temperature may also make Venus's lithosphere less brittle than Earth's colder rock. Again, if this hypothesis is correct, then Venus would have kept an Earth-like lithosphere and presumably ended up with plate tectonics if the runaway greenhouse effect had not driven temperatures so high.

A RECIPE FOR PLANETARY SUCCESS

We can now take everything we've learned to make a brief list of the things that make Earth a truly great planet, by which I mean a planet on which life has been able to evolve into intelligent beings. Let's start with the "surface-level" requirements:

1. A distance from the Sun that is great enough to allow water vapor to condense as rain and make oceans, but not so far that the water freezes.
2. Volcanism that can release trapped gases from the interior, including water vapor and carbon dioxide, to make an atmosphere.
3. Plate tectonics that can support a climate-regulating carbon dioxide cycle.
4. A fast enough rotation rate, combined with a core layer of convecting, molten metal, to generate a planetary magnetic field that protects the atmosphere from the solar wind.

These surface-level requirements might seem fairly stringent. However, if we are correct in guessing that Venus fails to satisfy #3 and #4 only because of its runaway greenhouse effect—which is a consequence of having formed a little too close to the Sun—then the recipe for planetary success might boil down to something as simple as this: *Assemble one planet from rock and metal within a star's habitable zone, and make sure it is at least within a few percent of being as large as Earth.*

In other words, any planet born close to Earth-*size* within its star's habitable zone might be expected to be Earth-*like* as well. This is a remarkable idea, because the example of Venus and Earth suggests that it's pretty likely that worlds of the right size get built. After all, if two out of the four planets in the inner region of our own solar system "happened" to end up at this size, we might expect planets of similar size to be common among other star systems. Combined with the fact that the habitable zone is at least moderately large, we reach the astonishing conclusion that unless we are misunderstanding some fundamental ideas, we would expect Earth-like planets to be quite common in the universe.

In fairness, there are some scientists who would say that I have indeed left out some important considerations, and that as a result planets like

Earth will prove to be quite rare. I'll discuss these "rare Earth" ideas in chapter 8, but I think you already know where I'm going to place my own bets. I suspect that Earth was destined for greatness by virtue of its size and location alone, and that it is a destiny shared by many other worlds.

MESSING WITH SUCCESS

We've covered the key ideas that are relevant to the prospects for finding other Earth-like planets that might give rise to intelligent beings and civilizations. But another obvious factor in the possibility of making contact with other civilizations is the survivability of those civilizations. As we discussed in chapter 3, even if civilizations arise fairly commonly, on average they would arise only thousands to tens of thousands of years apart in our galaxy. Thus, if civilizations typically destroy themselves within just decades or centuries after building their first radio transmitters or spacecraft, there may not be anyone out there for us to talk to.

Is self-destruction inevitable? Just a couple of decades ago, the prospect of nuclear annihilation seemed all too close. Today, it seems far less likely, though personally I think we'd be much safer still if we worked toward what we promised decades ago, and got rid of all our American nuclear bombs while coercing the rest of the world into doing the same. I know; there's the argument that someone, perhaps a terrorist group or a rogue state, will hide a bomb, so we should keep some of our own to deter their threat. But I just don't see it: Whatever you may think of our current military adventures, there's no doubt that the United States military tries hard to limit civilian casualties even when using conventional weapons. Given that, I don't see us wiping out hundreds of thousands of innocent civilians in order to strike back at some rogue group. The far better plan is to lower the likelihood of a nuclear attack in the first place by forcing all states to give up their weapons and to destroy the facilities that could make them, a plan more likely to be accepted if we said we'd willingly give up our own nuclear capability as well.

In any event, while nuclear annihilation will remain a threat for as long as the world maintains huge arsenals of weapons, I believe it has now been replaced at the top of the threat charts by a much more insidious concern: global warming. Global warming is insidious because it creeps up on you in a way that makes it easy to dismiss until it is too late. Even now, as evidence of an impending crisis mounts day by day, there are "nonbelievers" who think it's really not so bad, or even a hoax perpetrated by anti-capitalist environmentalists. But listen, folks: As you should realize after having read this chapter, the basic science really isn't very complicated. People can and do argue about the details, such as the feedback processes that may either mitigate or strengthen the warming over the short term and the precise effects of a warming climate on life and our civilization. These details are important to understand, but they don't change the fact that the basic threat is understood to a level far beyond reasonable doubt. Let's review.

Global warming is what we expect to occur if we humans continue to pour carbon dioxide and other greenhouse gases into our atmosphere. We expect this to warm our world because we already *know* that the naturally occurring greenhouse effect is the only thing that warms our world enough to keep it from being frozen over. Thus, if we strengthen the greenhouse effect further, we should expect the world to warm up even more. Further proof comes from Venus, where we *know* that a far stronger greenhouse effect turns the planet into a veritable hell. Given these things that we know, let me ask: Is there any possibility at all that we could keep dumping greenhouse gases into the atmosphere *without* it eventually causing our planet to warm up? No. Period.

For the short term, the only mitigating possibility is that feedback processes might delay the eventual warming for a while, but the evidence shows that this isn't happening. Although a few people still dispute it, measurements show clearly that Earth is indeed warming up quite rapidly by geological standards, that polar ice is indeed melting and raising sea level, and that severe storms are increasing in number and intensity due to the extra energy available as the oceans and atmosphere get warmer. And even if the minority view were correct, such that these seemingly clear measurements are being misinterpreted at present, the basic physics still wouldn't change: Add greenhouse gases and you'll add heat to your planet. Period. Again.

We live on a planet that is an incredible success, with a self-regulating climate mechanism that has made our existence possible. But we are now messing with success in a serious way. If you want to know how serious, forget all the debates about the rapidity of the warming and of the melting of the ice caps and all that other stuff, and just take a look at the graph in figure 6.6. It shows the change in atmospheric carbon dioxide concentration over the past 400,000 years, based primarily on ice core data for most of this period and on direct measurements for the past few decades. More recently acquired data extend the record back to about a million years ago and show the same basic thing: The carbon dioxide concentration has naturally

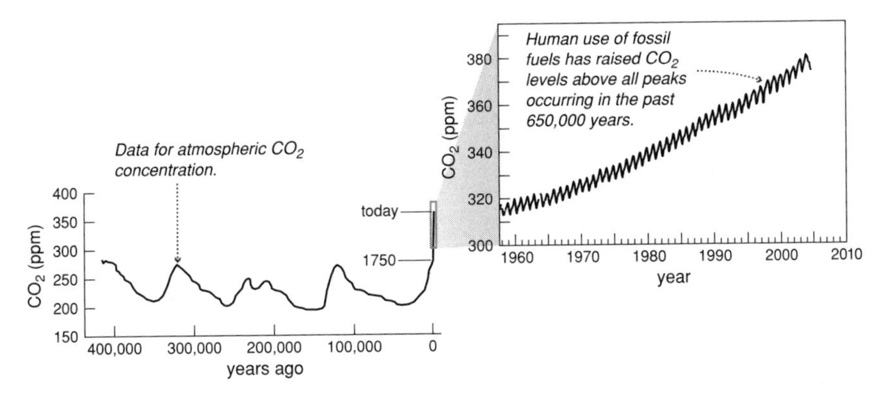

Figure 6.6. The carbon dioxide concentration in Earth's atmosphere over the past 400,000 years. Most of the data come from studies of air bubbles trapped in Antarctic ice (ice core samples); the data for the past half century (in the zoom-out) come from direct measurements made at Mauna Loa in Hawaii. (Illustration courtesy of Addison Wesley, an imprint of Pearson Education; data provided by the Carbon Dioxide Information Analysis Center)

fluctuated, but for at least the past million years it has never reached anywhere near what we have made it today. You'd have to have your head pretty deeply buried in the sand to look at these data and say there's nothing to worry about.

BEYOND UFOS

In the two previous chapters, we investigated the nature of life and the origin of life, seeing why we might be fairly confident about the prospects of finding microbial life on many other worlds. But we have seen that going from microbial to intelligent life—that is, to the types of aliens that might fly around in UFOs—is a more difficult step, and one that can happen only on a much more limited set of worlds. Earth has made the grade because it is of the right size and within the right range of distances from the Sun. If that's all there is to it, and I think it might be, then Earth-like planets should be common, greatly increasing the odds that other civilizations are out there.

Of course, we've layered a lot of reasonable but unproven assumptions upon one another. We've assumed that the rapid origin of life on Earth means that life would arise equally quickly on other similar worlds. We've assumed that the dinosaurs really did give way to mammals as a result of a random impact. We've assumed that Venus would have had plate tectonics and climate regulation without its runaway greenhouse effect. And so on. In the future, as we collect more data, conduct more experiments, and build better computer models of planets and climates, we should learn much more. But could we ever know *for sure* how all these events unfolded?

Probably not, because we can only read the few clues left behind through time. But I'll let you in on a little fantasy: You know how in Carl Sagan's *Contact*, the first message we receive from the stars is actually a transmission of one of our own early TV broadcasts, being beamed back to us from the star where it was received? Well, those aliens got the message because they were monitoring Earth, waiting for this type of radio signal to reach their receivers. But what if instead of monitoring Earth and our solar system for radio signals, they were just plain monitoring? Imagine that, a few billion years ago, a civilization that recognized the young Earth's future promise started a program to monitor our world (and perhaps Venus and Mars as well), in essence capturing continuous and incredibly high resolution video of our entire past. If so, and if we ever meet them and they let us search their archives, we might someday be able to see the events of Earth's distant past. Come to think of it, if they visited Earth and collected samples, they might even still have some dinosaurs living in their zoos. Likely? No. But way back in chapter 1, I promised you a book about possibilities. . . .

7

LIFE IN THE SOLAR SYSTEM

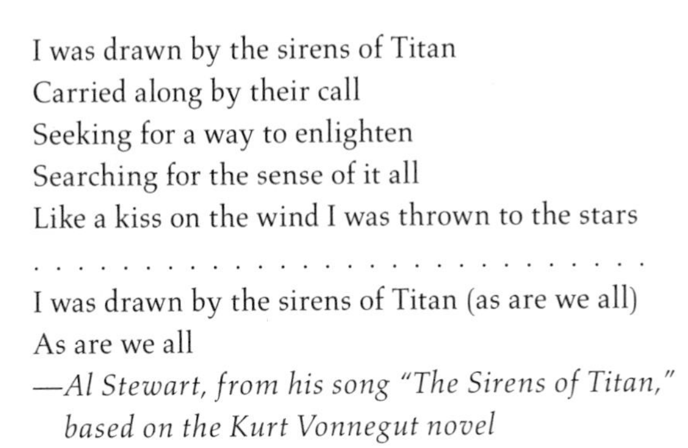

I was drawn by the sirens of Titan
Carried along by their call
Seeking for a way to enlighten
Searching for the sense of it all
Like a kiss on the wind I was thrown to the stars
. . . .
I was drawn by the sirens of Titan (as are we all)
As are we all
—*Al Stewart, from his song "The Sirens of Titan," based on the Kurt Vonnegut novel*

There's no place like home, at least here in our own solar system. There's no other world in our solar system on which we could survive even a few minutes outside without a spacesuit, and no other world that has so much as a puddle of liquid water on its surface. As I've explained over the past three chapters, this latter fact almost certainly means that we'll find no other intelligent life in our solar system, because surface water seems a clear requirement for the evolution of complex beings like ourselves.

But a lack of intelligent life does not necessarily mean a lack of all life, and, remarkably, there may be a half-dozen or more worlds in our solar system that offer at least the potential of harboring some type of life. If life actually exists on any of these worlds, it will not only give us the opportunity to learn much more about the general nature of biology, but also change the equation for the search for life elsewhere. If life happened more than once in just our own single solar system, then it seems inevitable that it has also happened on countless worlds around other stars.

Now, a half-dozen possible homes to life may seem a lot in one sense, but if you're a science fiction fan it might seem ridiculously limited. After all, there are a lot more "worlds" than that in our solar system. There are the 8 or 10 or more planets, depending on whether you count objects like

Pluto, Eris (the object discovered in 2005 that is slightly larger than Pluto), and other larger-than-Pluto worlds that may still await discovery. There are more than 150 known moons. And there are millions of rocky asteroids and perhaps a trillion icy comets. You can probably find science fiction stories that have placed life, perhaps even intelligent life, on almost every one of these myriad worlds.

As a science fiction fan myself, I enjoy these types of stories, but I suspect they are far more fiction than science. Based on what we know about the general nature of life, it just doesn't seem reasonable to have life without at least some type of liquid medium. So while I'm willing to admit possibilities beyond liquid water, such as liquid ammonia or liquid methane, I'm personally going to brush off the chance that we'll find life on any world that is solid throughout. This has the self-serving benefit of ruling out almost all the myriad worlds of our solar system—thereby allowing me to keep this chapter short—because they are almost all so small and so far from the Sun that they are frozen solid.

After we throw out all the small worlds as possible homes to life, we are left with just the planets and a few relatively large moons. I can brush off some of these pretty quickly too. I already dispatched Mercury in the prior chapter, because it suffers the same small size problems as our Moon and is too close to the Sun to boot. Venus, well, Venus is an interesting case. On the surface (literally), we can rule out life because of the high temperature that is caused by the runaway greenhouse effect. But remember that Venus *might* have had oceans in its early days, when the Sun was dimmer, and if so it is plausible to imagine that life may have arisen there. This possibility has led a few scientists to speculate that Venusian life may have adapted to the changing environment and taken shelter in the only conceivable niche left to it—the acidic droplets of high-altitude clouds. Frankly, I find this to be a stretch, but it's worth checking out. The European *Venus Express* mission, which arrived in Venus orbit in 2006, has instruments to probe the atmosphere that have at least some chance of detecting life among the clouds, if it exists. But beyond that, there's not much more I can say about the prospects for life on Venus, so I'll move on.

Let's next tackle the four large planets of the outer solar system—Jupiter, Saturn, Uranus, and Neptune. These planets are so different from the planets of the inner solar system that we use names to distinguish the two groups. The inner solar system planets—Mercury, Venus, Earth, and Mars—are made almost entirely of rock and metal, and we call them the *terrestrial* (Earth-like) planets because they are all Earth-like in composition, if not in other attributes. In an analogous way, we refer to Jupiter,

Saturn, Uranus, and Neptune as *jovian* (Jupiter-like) planets because they all resemble Jupiter in composition. These planets are all far more massive and larger in size than the terrestrial planets. Aside from metal and rock buried deep in their interiors, they are made mostly of hydrogen, helium, and hydrogen compounds such as water (H_2O), methane (CH_4), and ammonia (NH_3). Although they are sometimes called "gas giants" because the materials that make them are generally gaseous under terrestrial conditions, in reality the jovian planets have only relatively thin outer layers of gas. Beneath that, the internal pressures and temperatures become so extreme that the "gases" are compressed into liquid form, or even into other phases that we never see under natural conditions on Earth. In other words, there's no place to "land" on any of these worlds; if you descended into one, you could continue downward until you were crushed by the growing pressure, at which point your remains would sink until they settled in at a depth at which the density would allow them to float.

Could the jovian planets have life? There are two places where they might have the necessary liquids. First, some of their clouds have droplets of liquid water (and other liquids as well). However, life in these clouds seems highly unlikely, because these planets also have very strong vertical winds with speeds that would make a hurricane seem like a gentle breeze. Any complex organic molecules that might assemble would quickly be carried to depths at which the heat would destroy them, making it difficult to see how life could arise. The only way that anyone has imagined life surviving in jovian clouds is by supposing that it might have some sort of buoyancy that allows it to stay at the right altitude while the vertical winds rush by it. However, such buoyancy would require large gas-filled sacs, making the organisms themselves enormous. Given that we cannot envision a way for microbes to survive, there seems to be no way for large, buoyant organisms to evolve to begin with.

The second place where jovian planets might conceivably have life is deep in their interiors. Jupiter and Saturn probably lack any liquid layer aside from liquid hydrogen, which does not seem a suitable medium for life, but Uranus and Neptune offer another possibility. Theoretical models suggest that these two planets should have deeply buried core layers consisting of a liquid mixture of water, methane, and ammonia (sitting atop a central core of metal and rock), making for very odd "oceans." The high pressures, strange mix of liquids, and a lack of any obvious way to extract energy from these "oceans" makes life within them seem highly unlikely, and even if life were possible we lack any technology that could allow us to search for it in the cores of these giant planets. As a result, I'll have no more to say about the possibility of life on jovian planets.

To review where we stand, I've ruled out all the small worlds of our solar system as homes to life because they lack any liquids at all. I've dropped the Moon and Mercury for the same basic reason. I've allowed a slim possibility for life in the clouds of Venus or in deeply buried core layers of Uranus and Neptune. That leaves us with the rest of this chapter, in which I'll take you on a quick tour of the remaining worlds—that is, those worlds that seem like potential candidates for life. We'll start with Mars, which has long been the favorite alien home of science fiction, and happens also to be the world that most scientists would name as "most likely to have life" (aside from Earth) in our solar system.

MARS

Have you ever noticed that people often speak of Martians but rarely of, say, Venusians or Jupiterians? It's no accident, but rather the result of misconceptions that arose from early telescopic observations of Mars.

The story begins in the late eighteenth century with the brother and sister astronomers William and Caroline Herschel. Although they are most famous for discovering the planet Uranus, they also made many observations of Mars. Their observations revealed several uncanny resemblances to Earth, including a similar axis tilt, a day just slightly longer than 24 hours, the presence of polar caps, and seasonal variations in appearance over the course of the Martian year (about 1.9 Earth years). The Herschels assumed that Mars must be quite Earth-like, and even speculated about the nature of its inhabitants.

The hypothetical Martians got a bigger break about a century later. In 1879, Italian astronomer Giovanni Schiaparelli reported seeing a network of linear features on Mars that he named *canali,* by which he meant the Italian word for "channels" but which was frequently translated as "canals." Coming amid the excitement that followed the 1869 opening of the Suez Canal, Schiaparelli's discovery soon inspired visions of artificial waterways built by an advanced civilization. The reports captured the imagination of American astronomer Percival Lowell, who commissioned the building of an observatory for the study of Mars in Flagstaff, Arizona.

The Lowell Observatory opened in 1894. Barely a year later, Lowell published detailed maps of the Martian canals and the first of three books in which he argued that the canals were the work of an advanced civilization. He suggested that Mars was falling victim to unfavorable climate changes and that the canals had been built to carry water from the poles to thirsty cities elsewhere. Lowell's work drove rampant speculation about Martians and fueled science fiction fantasies such as H. G. Wells's *The War of the Worlds*, published in 1898.

Given that the canals don't actually exist, what was Lowell seeing? In a few cases, his canals correspond to real features on Mars. For example, the canal he claimed to see most often (which he called *Agathodaemon*) coincides with the location of the huge canyon network now known as Valles Marineris (see the map of Mars in color plate 4). For the most part, however, Lowell must simply have imagined the canals, perhaps offering an extreme example of the human tendency to see vivid patterns where none really exist, much as we see patterns in ink blots or among the stars that speckle our sky.

Lowell was well known both to scientists and to the public, so his claims generated a lot of attention. Nevertheless, most other scientists remained skeptical, because they could not see the canals through their own telescopes and they did not show up in photographs. Some also found flaws in Lowell's basic assumptions. For example, Alfred Russel Wallace, more famous for having developed the theory of evolution by natural selection independently of but at about the same time as Darwin, used physical arguments to conclude that Mars must be too cold for liquid water to flow on its surface. Moreover, Wallace pointed out that Lowell's canals followed straight-line paths for hundreds or thousands of miles, while real canals would be expected to follow natural contours of topography (for example, to go around mountains). On this point, Wallace wrote that the canals described by Lowell "would be the work of a body of madmen rather than of intelligent beings."

Lowell's story illustrates both the pitfalls and the triumphs of modern science. The pitfall is that individual scientists, no matter how upstanding and dedicated, may still bring personal biases to bear on their scientific work. In Lowell's case, he was so convinced of the existence of canals and Martians that he simply ignored all evidence to the contrary. But the story's ending shows why modern science ultimately is so successful. Despite Lowell's stature, other scientists did not accept his claims on faith. Instead, they sought to confirm his observations and to test his underlying assumptions. They found that Lowell's claims fell short on all counts. As a result, Lowell became an increasingly isolated voice as he continued to advocate a viewpoint that was clearly wrong.

The story also illustrates the divide that often persists between scientists and the general public. Although nearly all scientists had abandoned belief in Martians within the first few decades of the twentieth century, the canal myth persisted among the public. Belief in Martians remained widespread enough to create a panic during Orson Welles's 1938 radio broadcast of *The War of the Worlds*, when many people thought a Martian invasion was actually underway. Even today, a few die-hard proponents still claim to see evidence of a past Martian civilization, but the features they "see"—such as the famous "face on Mars"—are no more real than Lowell's canals. We now have high-resolution photographs of the entire Martian surface. There are no canals, no faces, and no ruins.

Nevertheless, as I described in chapter 6, there is abundant evidence that water once flowed on the surface of Mars. This fact alone gives us reason to imagine that life could have arisen on Mars, or survived there if it had arrived on meteorites from Earth. That is the primary reason why Mars has been and will continue to be the target of more space missions than any other planet.

Our first close-up photographs of Mars arrived in 1965, beamed back by radio waves from NASA's *Mariner 4* spacecraft as it passed within about 6,000 miles of the Martian surface. *Mariner 4* captured 22 low-resolution photographs showing just 1 percent of the planet during its brief close encounter. By chance, the pictures turned out to be of some of the more heavily cratered regions of Mars, so they gave a somewhat misleading first impression of Mars. In 1971, the *Mariner 9* spacecraft went into orbit of Mars, and thereby was able to take enough pictures to give us a better global view of the planet. These photos made it abundantly clear that Mars was not a good place for a civilization, but they also began to show the features carved by running water in the distant past. Our views of Mars improved considerably more in 1976, when two *Viking* spacecraft arrived, each with an orbiter to photograph the planet from above and a lander to drop down to the surface.

The *Viking* orbiters provided clear, photographic evidence of abundant water in the Martian past, and at the same time the *Viking* landers conducted the first robotic experiments designed to search for life on Mars. Each lander was equipped with a robotic arm to scoop up a bit of Martian soil, and the soil was then placed into chambers where it was subjected to heating, mixing, and various other simple experiments. Although these experiments offered some intriguing results that at first made some scientists think they might have detected life on Mars, the soil showed no sign of containing any organic molecules *at all*. This lack of organic molecules implies that the intriguing experimental results were almost certainly due to chemical rather than biological reactions, which is why all but a tiny minority of scientists now say that the *Viking* results were negative for life.

The *Viking* missions provided a wealth of scientific data about Mars. But they also left many questions unanswered, and the scientific community was itching for follow-up missions. Unfortunately, budgetary and political considerations, along with the failure of two Russian missions to Mars (*Phobos 1* and *2*) and one American mission (*Mars Observer*), all conspired to stop spacecraft exploration of Mars for some twenty years. The long mission drought did not end until July 4, 1997, with the landing of *Pathfinder* and its little rover, *Sojourner*, named for Sojourner Truth, an African American heroine of the Civil War era. Although *Sojourner* could travel only a few tens of yards from *Pathfinder*, this was enough to check the chemical composition of many nearby rocks. Around the same time, the *Mars Global Surveyor* arrived in Martian orbit, from which it took high-resolution surface photographs for nine years.

The *Pathfinder* and *Mars Global Surveyor* missions marked the beginning of a new era of intensive scientific study of Mars, and also marked a change in strategy from the *Viking* era. As scientists learned more both about Mars and about the nature of life, it became clear that we are not quite ready to undertake a serious search for life. Because Mars has no liquid water on its surface today, any extant life would presumably be underground at depths where internal heat can keep water liquid. In other words, Martian life today would probably resemble the terrestrial microbes known as *endoliths* that live in subsurface rock on Earth. Searching for Martian life therefore presents several difficult challenges: We'd need to drill down to bring up rock from fairly deep underground, we'd need to do that at a location where a heat source is keeping some of the water liquid, and then we'd need to conduct careful experiments to detect the presence of microscopic life. As a result, scientists decided that for the moment, it makes better sense to spend some time learning more about Mars generally, so that when we later undertake a search for life we'll be able to choose the best places to look and to employ the best strategies for detection.

With that in mind, scientists developed a step-by-step strategy for studying Mars in which each new mission builds upon the results of prior ones. This strategy keeps costs down and allows us to employ new technologies as they come along rather than trying to do everything at once with existing technology. To date, it's been a very successful strategy. With current rockets, we can send missions to Mars only around the times that it lines up with Earth in its orbit around the Sun (because it is too far away at other times), and these launch opportunities occur about every 26 months. Although the next attempted missions after *Pathfinder* and *Mars Global Surveyor* failed in 1999, we've since had a series of successes. The *Mars Odyssey* orbiter arrived in 2001; the *Spirit* and *Opportunity* rovers, along with the European *Mars Express* orbiter, in early 2004; and the *Mars Reconnaissance Orbiter* in 2006. The *Phoenix* lander should set down on Mars in May 2008, shortly after this book is published. Additional missions are being planned for each of the upcoming future launch opportunities, including the *Mars Science Laboratory*, a rover scheduled to land on Mars in 2010.

These missions are unlikely to turn up direct evidence of life, but they have already provided tantalizing hints that Mars was habitable in the past and may still have a habitable subsurface today. For example, *Mars Odyssey* has shown that subsurface water ice is abundant on much of Mars—which also means that there could be subsurface liquid water in any place that enough heat is available to melt the ice. *Mars Express* and the *Mars Reconnaissance Orbiter* have been transmitting images of stunning clarity, showing the paths of ancient rivers, layering of ice at the polar caps, and possible evidence of ancient oceans.

Perhaps most spectacularly, the *Spirit* and *Opportunity* rovers, each designed to last only three months, are still going strong as I write this book, well over three years after their arrival. They have provided us with startling vistas of the Martian surface, as well as close-up studies of Martian rocks. Both landing sites offer further evidence of past liquid water. For example, rocks at the *Opportunity* landing site contain tiny spheres—nicknamed "blueberries" although they're neither blue nor as large as the berries we find in stores—and odd indentations suggesting that they formed in standing water (color plate 5). Compositional analysis shows that the abundant "blueberries" contain the iron-rich mineral hematite, and other rocks contain the sulfur-rich mineral jarosite. Both of these minerals form in water, and chemical analysis seems to support the case for formation in a salty environment such as a sea or ocean. The layering observed in the sedimentary rocks further supports the case for a past sea or ocean at the landing site.

Meanwhile, orbital photographs suggest that although the global wet period ended billions of years ago, at least some regional water flows may have occurred intermittently ever since. This is not as far-fetched as it may sound. Although liquid water is unstable on the surface of Mars today, a large quantity of water that suddenly erupts onto the landscape will take a little time to freeze or evaporate completely. A catastrophic release of

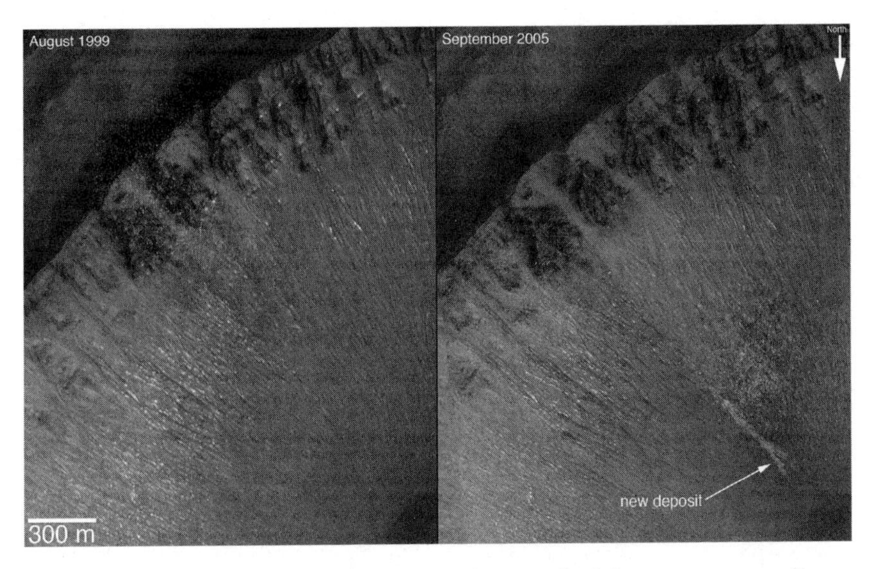

Figure 7.1. The *Mars Global Surveyor* photographed the same crater wall twice, six years apart. Notice that the 2005 photo shows a new gully that was not present in the 1999 photo. (Courtesy NASA/JPL-Caltech)

floodwaters—from beneath the surface, for example—could survive long enough to carve channels and other surface features. We do indeed see surface features that indicate periodic flooding over the past few billion years.

More spectacularly, comparison photos taken a few years apart strongly suggest that at least some small-scale water flows still occur. Figure 7.1 shows orbital photos of the same crater wall, taken six years apart by *Mars Global Surveyor*. The many gullies look strikingly similar to the gullies we see on almost any eroded slope on Earth, and they probably form when subsurface ice melts and gushes out from the crater walls before rapidly freezing or evaporating in the thin Martian atmosphere. Notice that at least one new gully appears to have formed between the times the two photos were taken.

Theoretical modeling is also painting a new picture of Martian history. Earth and Mars both have nearly the same axis tilt today, and hence similar seasonal patterns, but this is apparently a coincidence of the moment. A planet's axis tilt can change with time due to the gravitational influence of other planets, such as Jupiter. Earth's axis tilt changes very little with time—varying only between about 22° and 25°—because our large moon exerts a gravitational pull that stabilizes it. Even so, these small changes in tilt have been linked to Earth's cycles of ice ages, because when the tilt is smaller the poles get less summer sunlight, and the lesser sunlight allows more water to freeze and glaciers to advance. Mars lacks a large moon, and its two tiny moons (Phobos and Deimos) are far too small to offer any stabilizing influence on its axis. In addition, because Mars is closer than Earth to Jupiter, Jupiter's gravity more strongly perturbs Mars as it orbits the Sun. Calculations suggest that, together, the lack of a stabilizing moon and the effects of Jupiter should cause Mars to experience wild swings in its axis tilt—taking it anywhere between 0° and about 80°—on time scales of a few hundred thousand years.

These changes in tilt would have dramatic effects on the Martian climate. When the axis tilt is small, the poles may stay in a perpetual deep freeze for tens of thousands of years. With more carbon dioxide frozen at the poles, the atmosphere becomes thinner, lowering the pressure and weakening the greenhouse effect. In contrast, when the axis is highly tilted, the summer pole should become warm enough to allow substantial amounts of water ice to sublimate into the atmosphere, along with all the carbon dioxide. The atmospheric pressure therefore increases, and Mars becomes warmer as the greenhouse effect strengthens. The Martian polar regions show layering that probably reflects changes in climate due to the changing axis tilt. The atmospheric pressure probably does not become high enough to allow liquid water to pool in surface lakes or ponds, but it may allow ice to melt into liquid water just beneath the surface, and occasionally to gush onto the surface for short periods of time before it freezes or evaporates.

Now, I know all this may seem like overwhelming detail, and I don't even expect my students to memorize all this stuff. But I've told it to you so that you'll understand that scientists are no longer imagining things the way Percival Lowell once did. Instead, we are using real, reproducible evidence from multiple sources, and thereby learning what Mars really is like today and has been like in the past. So let's briefly review the big picture of Mars as we now understand it.

Mars was born at the same time as Earth, about 4 ½ billion years ago. Early in its history, it had volcanic outgassing that produced an atmosphere much thicker and with a much stronger greenhouse effect than the atmosphere it has today. The atmosphere was warm enough and dense enough to allow water vapor to condense as rain, and there was a time when rivers flowed and lakes and perhaps even oceans filled. Although this global wet period ended by some 2–3 billion years ago, lots of water ice still exists on and beneath the surface of Mars, and it sometimes melts and emerges to flow briefly across the surface.

This simple summary of Martian history should make the implications for life clear. Early in its history, Mars seems to have had everything that Earth had, so if life got started here, why not there? And even if life on Mars did not arise indigenously, microbes from Earth almost certainly arrived by meteorite, and if they landed in pools of water they may well have survived. So unless we've got some critical element of this story wrong, it seems that Mars *should* have had life in the past. Given the evidence of abundant subsurface ice and that water sometimes still flows, it seems reasonable to infer that at least some underground locations have had liquid water continuously over the past few billion years, which means that if Mars had life in the past, it could have survived all the way to the present. Perhaps now you understand why most scientists would vote Mars "most likely to have life."

Of course, we won't really know if Mars has ever had life until we find living organisms or fossils of past life,[1] or until we've explored the planet thoroughly enough to conclude that no life exists. But at least we now know what we're looking for.

Scientists argue amongst themselves about whether it would be better to search for life on Mars with robots or people, but I don't consider it a scientific question. Instead, I see it as a sociological question, and if you've read my children's books, you know where I stand. I believe that sending people back to the Moon, and eventually on to Mars, could bring people together in a way that *nothing* else could accomplish, inspiring children to work hard toward a world where everyone lives in peace and shares in the advance of human knowledge. Robotic science is cool, but no one grows up with the dream of being a robot. Inspiration comes from people.

So with luck, maybe in two or three decades, the first human explorers will be reaching Mars. What will they find? Perhaps, while chipping out rock from the walls of Valles Marineris, someone will discover a pebble containing microfossils of past life on Mars. Or maybe, walking along the

[1] Some of you are probably wondering why I haven't devoted any space to the claims of fossil evidence for life in a Martian meteorite (ALH84001). Two reasons: First, because I discuss them in some depth in my earlier book, *On the Cosmic Horizon*, and in even more depth in my textbooks. More important, it's because while I personally find the claims intriguing, I doubt that they can be substantiated until we study rocks collected *on* Mars; the problem with meteorites found on Earth is that it's too difficult to rule out terrestrial contamination of evidence.

northern plains, someone will stumble across a stromatolite, formed by a microbial colony that lived in ancient Martian seas. My friend Alan Stern, leader of the *New Horizons* mission to Pluto (which launched in 2006 and will reach Pluto in 2015) and recently appointed as NASA's head of space science, suggests that we might even find fossils of larger organisms, such as fossil seashells. It's a plausible idea if Mars was continuously wet for a long enough period to allow evolution to progress that far. But what I'm really hoping for is actual, living life. If I had to place a bet, I'd bet that it's there, just waiting to be discovered when, as happens in my children's book *Max Goes to Mars,* we drill down into the ground under a dry riverbed, near a not-quite-dead Martian volcano, and find the indisputable proof that we are not alone in the universe.

So look around you, at the young children you see at the park and the school playground, and at the babies coming out of the maternity wards, and don't forget this: One of those little people is going to be the first person to walk on Mars, and perhaps the person who answers once and for all the question of whether life exists beyond Earth. I, for one, can hardly wait.

THE GALILEAN MOONS OF JUPITER

The next stop on our biological tour is Jupiter, or at least its four largest moons: Io, Europa, Ganymede, and Callisto. These moons were discovered by Galileo in 1610, shortly after he first turned his newly built telescope to the heavens and began making the epic discoveries that helped seal the triumph of the Copernican revolution. In the historical context of the times, the discovery of these four *Galilean moons* was just one of several critical observations by Galileo that contributed to the downfall of the geocentric model after Kepler published his laws of planetary motion. But while their existence did not by itself prove that Earth was orbiting the Sun, it made one thing very clear: Since these moons were orbiting Jupiter, the Earth could no longer be considered the center of everything.

The Galilean moons are large enough that we would probably consider them planets if they orbited the Sun rather than Jupiter. Ganymede, which is the largest moon in the solar system, is larger than the planet Mercury. The other three all are considerably larger than Pluto. Besides being planet-like in size, they are planet-like in character, with warm interiors and surfaces on which geology has clearly operated.

Visually, Io is the most spectacular. It is the innermost of the four moons, and has the distinction of being by far the most volcanically active world in the solar system. Its entire surface is pockmarked by large, active volcanoes, and at least some are erupting at almost all times. This incredible activity pretty much rules out life on Io, since any living organisms would soon be buried in molten lava. However, it raises an important question: Although Io is big compared to most moons, it is only slightly larger in size than our own cold, dead Moon; so how can it possibly retain enough internal heat to power so many volcanoes?

The answer lies with a remarkable heating mechanism known as *tidal heating*. The three innermost Galilean moons—Io, Europa, and Ganymede—share an interesting relationship in their orbits of Jupiter: During the time it takes Ganymede to orbit Jupiter once (about seven days), Europa orbits exactly twice and Io orbits exactly four times. This "orbital resonance" means the moons periodically fall into a straight line, and their mutual gravitational pulls therefore make their orbits more elliptical than they otherwise would be. For reasons I won't go into here, the ellipticity of the orbits causes the moons to feel a varying tidal pull from Jupiter, one that in essence churns their insides back and forth, generating friction and heat. Io experiences the most tidal heating because it is closest to Jupiter, and that is why it is hot enough inside to be so volcanically active.

The orbital resonance that makes the tidal heating possible might seem like an incredible coincidence, but it is not. Just as our own Moon's synchronous rotation—meaning the fact that it always keeps the same face toward Earth—is a natural consequence of its gravitational interactions with Earth, orbital resonances tend to arise wherever multiple moons orbit a large planet in fairly close proximity. Indeed, these Galilean moons are not the only cases in which we find orbital resonances among moons in our solar system. Similar resonances also affect the asteroids of the asteroid belt, Pluto and other similar objects of the outer solar system, and the boulders orbiting in the rings of Saturn. We've even discovered resonances among some of the planets that orbit other stars.

Tidal heating is clearly overdone on Io from the standpoint of life, but it may be "just right" on the next moon out, Europa. Europa has a remarkably smooth surface, with very few impact craters and no large mountains (figure 7.2). It is also criss-crossed by large cracks. Cracks in what? In fact, Europa's entire surface is made of *water ice*, and closer-up photos show not only the large cracks, but ample evidence that water, or at least slushy ice, occasionally breaks through from underground. The implication is

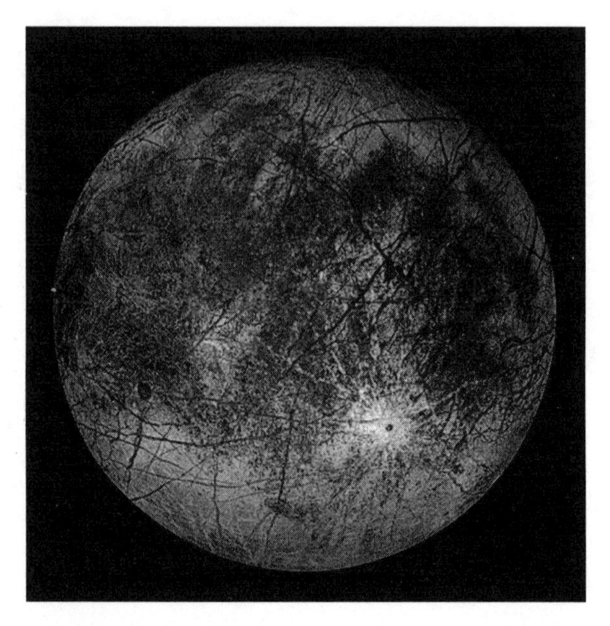

Figure 7.2. A global view of Europa, as seen from the *Galileo* spacecraft, which orbited Jupiter between 1995 and 2003.

probably obvious: There's either liquid water or ice that is at least partially melted somewhere under the Europan surface.

Models of Europa based on studies of its density and gravitational field, combined with calculations of the amount of tidal heating it receives, paint an even more astonishing picture. Europa is made mostly of rocky material, but it has a thick outer shell that is made out of H_2O. The outermost portion of this shell is frozen solid, perhaps down to a depth of 10–20 miles. But beneath that there *probably* lies an ocean of liquid water that is more than 50 miles deep. All told, Europa may have two to three times as much ocean water as Earth.

Now, you may rightly say that "surface photos and models are fine, but do we have any direct evidence that the ocean really exists?" The answer is "sort of." Until we land a heated spacecraft on Europa and watch what happens to it as it melts through miles of ice, we have no way of *seeing* whether liquid water really does lie under the ice. But there are other ways to learn whether there's an ocean underfoot, and one piece of evidence looks particularly strong: magnetic field data collected by the *Galileo* spacecraft. *Galileo* entered orbit of Jupiter in 1995, and it is the only orbiter we've ever sent to the largest planet in our solar system; all the other spacecraft that have visited Jupiter, including *Pioneer 10* and *11*, *Voyager 1* and *2*, and more recently *New Horizons*, were flybys that made only a quick pass as they flew outward through the solar system.

In 1996, *Galileo*'s magnetometer detected a magnetic field near Europa. But the field wasn't steady; instead, it varied in tune with Jupiter's approximately ten-hour rotation. Why would Europa's magnetic field be responding to Jupiter's spin? There's only one plausible explanation: Just as a moving magnet produces an electric current in a coil of wire, Jupiter's rotating magnetic field must be producing electrical currents within Europa that then generate an "induced" magnetic field. As far as we know, such electric currents could be produced within Europa only if it has a liquid layer, not if it is frozen throughout. Moreover, the particular character of the induced magnetic field is consistent with the idea that Europa has a deep, subsurface ocean of *salty* water.

If you add it all up, here's what we seem to have on Europa: a rocky interior kept warm by tidal heating, with a salty ocean above and a frozen outer crust. Given this picture, it's reasonable—but far from proven—to think that Europa might have active volcanism on its rocky seafloor. In other words, Europa may have deep-sea vents very much like those on Earth that we consider to be the most likely sites of the origin of life some 4 billion years ago. Thus, if our picture is correct, Europa may have the same deep-water conditions that gave rise to life on Earth, making life on Europa seem like at least a moderately good bet.

If there is life on Europa, what might it look like? Since we can't see through the icy crust, the possibilities might seem almost limitless. In my public talks, I used to tell audiences that there could be *whales* swimming in the Europan ocean and we'd have no way to know it. But one day, a ten-year-old in the audience (who happened to be a second cousin of mine) asked, "If there were whales, and the whole moon is covered with ice, how would they breathe?" Good point. So, instead, I'll tell you that there could be *really big fish* swimming in the Europan ocean, and we'd have no way to know it.

However, when we look at the possible energy sources for life, the big fish look a lot less likely. While deep-sea vents might offer enough energy for an origin of life, they could not by themselves support more than a small total biomass, because they simply don't make enough energy available to living organisms. This fact might surprise you if you think about the great communities of life that live near deep-sea vents on Earth today, but most of this life actually gets its energy from materials that filter down from above, such as dead organisms and oxygen produced by photosynthetic life near the surface. Only a small fraction of the life near Earth's deep-sea vents lives solely off energy from the vents themselves. For life to be abundant or widespread on Europa, it would need some other energy source in addition to the chemical reactions near deep-sea vents. Photosynthesis seems out, since sunlight cannot penetrate through the thick icy crust. And while scientists have considered a few other potential sources of energy for life on Europa, none of them seem to add up to anything close to the amount of energy that photosynthesis makes available in Earth's oceans. As a result, while life in the Europan oceans seems like a reasonable possibility, we doubt that there is enough energy to have allowed that life to evolve very far or to be particularly abundant.

The prospects for life on Europa make it an inviting target for future exploration. As budgets go up and down, NASA has for several years been working on and off on plans to send an orbiter to Europa. The orbiter could make detailed measurements intended to settle the case as to whether an ocean really exists. If the ocean proves real, the next mission might land on Europa, where, by melting and filtering some of the surface ice, it would have at least some chance of turning up evidence for life in the ocean below. In the meantime, scientists take the possibility of life on Europa so seriously that they deliberately ended the *Galileo* mission in 2003 by causing it to crash into Jupiter's atmosphere, thereby preventing any possibility that the spacecraft might someday crash into and contaminate Europa with hitchhiking microbes from Earth.

Continuing outward from Jupiter past Europa, we next encounter Ganymede. Because of its greater distance from Jupiter, Ganymede has less tidal heating than Europa. However, its larger size may mean it also retains more heat from radioactive decay, and the combination of radioactivity and tidal heat may be enough to give Ganymede a subsurface ocean as well. Photographs do indeed show a few features suggesting that liquid water has broken through to the surface of Ganymede in the past. In addition, Ganymede has an induced magnetic field much like that of Europa, again suggesting the presence of a subsurface, salty ocean.

While the possibility of a subsurface ocean is encouraging from the standpoint of habitability, the lesser heating on Ganymede means that the ice cover would be much thicker than on Europa—probably at least 100 miles thick. This would make finding life in a subsurface ocean far more difficult and the transport of possible nutrients or energy from the surface considerably less efficient. In addition, the pressure in Ganymede's interior is high enough to create high-density forms of ice that sink in water—

rather than floating like ordinary ice—and that therefore would lie between the rocky interior and any liquid water ocean. As a result, it is much less likely that Ganymede has seafloor vents or any rock-water interfaces that could facilitate the types of chemical reactions thought to be necessary for an origin of life. Even if life does exist, the available energy seems much more limited than on Europa, so we would expect it to be even simpler and less abundant.

The outermost of the Galilean moons, Callisto, does not participate in the orbital resonance of the other three and therefore has no tidal heating at all. Its surface is made of water ice, but it is densely cratered and shows no evidence of liquid water ever having broken through the ice. As a result, we wouldn't expect to find a subsurface ocean on Callisto—but, surprisingly, the magnetic field data say otherwise. Again, it has an induced magnetic field suggesting the presence of a salty ocean. Scientists are still struggling to understand how this ocean can exist (if indeed it does), but have some plausible ideas. For example, the nature of Callisto's icy crust may make it a very good insulator, thereby allowing this moon to retain internal heat better than most other worlds. In addition, the water may contain dissolved salts and ammonia that effectively act like antifreeze. The ocean would make Callisto another candidate for life. However, there's probably even less energy available for life on Callisto than on Ganymede, making it the longest shot for life of the three potentially habitable Galilean moons.

I've given you all the appropriate caveats about the energy for life on the Galilean moons, but let's ignore those for the moment. If we focus just on the positives, we have *three* moons that may each have more ocean water than Earth, all orbiting just one of the planets in our solar system. If they really have oceans, and if life is possible within these oceans, then we confront the astonishing prospect that Jupiter alone could be host to three distinct types of biology against Earth's one. In that case, alien biologists might learn more by doing comparative studies at Jupiter than by visiting Earth.

TITAN

Our biological tour next takes us to Saturn, where we encounter what may be the most amazing moon in the solar system: smog-covered Titan. Titan is the second largest moon after Ganymede, and like Ganymede it is larger than the planet Mercury. But Titan's real claim to fame—aside from having had a novel and song written for it (see the quotation at the beginning of this chapter), as well as a terrific movie (*Gattaca,* 1997) made about a boy who dreamed of visiting it—is that it has a thick atmosphere surrounding it.

Titan's atmosphere is even thicker than Earth's. The surface pressure is about 50 percent greater than that on Earth, which means that if you could visit Titan, the pressure would feel fairly comfortable even without a spacesuit. The temperature, however, would not. Here, where sunlight is nearly 100 times as weak as on Earth, the surface temperature is a frigid –290°F. Moreover, while the atmosphere is 90 percent nitrogen (N_2)—not so different from the 77 percent nitrogen content of Earth's atmosphere—there is no appreciable oxygen to breathe.

Our first close-up look at Titan came with the *Voyager* flybys in 1980 and 1981. These flybys showed us nothing of the surface, because Titan is completely enshrouded by a thick haze and clouds. Nevertheless, the *Voyagers* told us a lot about Titan. Titan's gravitational tug on the *Voyager* spacecraft allowed us to determine its mass and likely overall composition: Titan is made about half and half of rock and ice, but the ice includes not only water ice like the Galilean moons but also ammonia ice and methane ice. This makes sense, because Saturn is farther from the Sun than Jupiter, so its moons formed in a colder region of the solar system where ammonia and methane were able to condense along with water ice. In addition, instruments on the *Voyager* spacecraft measured Titan's atmospheric temperature and content. Results showed that after nitrogen, the next most abundant gases in Titan's atmosphere are methane (CH_4), argon (Ar), and ethane (C_2H_6), and there are lesser quantities of gases including propane (C_3H_8), acetylene (C_2H_2), and hydrogen cyanide (HCN). In other words, Titan's atmosphere is loaded with hydrocarbons, and this fact really got the scientists thinking.

Titan's surface is far too cold to have liquid water, but the mere presence of all these hydrocarbons suggested that Titan must have some very interesting prebiotic chemistry going on. That is, even if it has no life, the chemistry on Titan's surface might tell us a lot about the natural types of chemistry that could lead to life on a warmer world. In addition, while we can rule out liquid water, the *Voyager* results suggested that Titan should have rainfall made up of cold droplets of liquid methane and ethane: in essence, liquid natural gas. And if that wasn't already enough to make scientists say "we must go back there," get this: The atmospheric methane on Titan tells us that there must be a methane source on or below the surface, because methane wouldn't last long in the atmosphere on its own. Scientists guessed that the source might be liquid methane on the surface, suggesting that Titan might have lakes or oceans of liquid methane (and ethane). Could there be cold, methane-based life on Titan?

As I discussed in chapter 4, biologists suspect that liquid methane would not work as well as water as a liquid medium for life, and its low temperature would in any event imply much slower chemical reaction rates than would occur for similar reactions in water. If methane life could get started at all, it would probably have a very slow metabolism. Nevertheless, we really don't know enough to rule out methane life without a lot more study, and given the fact that Titan was an intriguing place anyway, it became a prime target for follow-up studies. So in 1997, NASA launched the two-ton *Cassini-Huygens* spacecraft to Saturn.

After a circuitous, seven-year journey that took it twice past Venus, back past Earth, and then on past Jupiter—with each planetary pass giving it a boost in speed—*Cassini* reached Saturn orbit in July 2004. With infrared cameras that can "see" through the smoggy atmosphere, *Cassini* quickly gave us much clearer views of Titan than we had ever had before. *Cassini* is scheduled to continue operating in Saturn orbit until at least mid-2008, by which time it will have made more than forty close passes of Titan, each time collecting more data and photographing different regions of the moon in greater detail. *Cassini* has already turned up many incredible sights, including dozens of methane-ethane lakes, features that look like "ice volcanoes" that sometimes erupt to release methane gas and an "icy lava," and wind-sculpted dunes that may be composed of organic hydrocarbons. Scientists are hoping that this rich science will allow the *Cassini* mission to be extended beyond its scheduled shutdown date.

This is all very impressive, but when I really want to be impressed with human ingenuity I think about this: Titan is nearly a billion miles away from Earth, and we've *landed* there. That's right; *Cassini* is an orbiter, but it didn't go to Saturn alone. Attached throughout its journey, it carried a European-built probe named *Huygens* (after the Dutch scientist Christiaan Huygens). On Christmas Day, 2004, Cassini released the probe, which then spent 21 days coasting through the space around Saturn as it approached Titan. On January 14, 2005, *Huygens* arrived at Titan, where it plunged into the atmosphere, deployed a series of parachutes, and, after a 2 ½-hour descent, landed on the surface of this distant world. If you like analogies, just hitting a moon the size of Titan from Earth is the equivalent of shooting a gun and hitting a dime from a distance of 2,500 miles away. Hitting it softly enough to land and take pictures of the surface . . . well, I can't think of any words that would do it justice.

So as you gaze at the surface of Titan (color plate 6), I hope you'll take it as a lesson not only about the possibility of life in our solar system, but about what human beings are capable of when we use our creative powers to build rather than to destroy. It may sound corny, but seeing chunks of ice littering the landscape of Titan gives me hope for the human race.

Now back to the science. During its descent, instruments aboard *Huygens* studied Titan's atmosphere while cameras snapped hundreds of photos of the approaching landscape below. These aerial views showed fantastic scenery, including river valleys that merge together and flow down toward what appears to be the shoreline of a lake (see color plate 6). The rivers and lake are dry now, but it seems almost certain that cold rain does indeed sometimes fall on Titan, occasionally filling rivers and lakes with liquid methane. *Huygens* hit the surface fairly gently, at a speed of about 10 miles per hour, where it operated, as planned, for about 90 minutes. The landing site looks much like a dry streambed, strewn with boulders that are actually granite-hard chunks of ice, probably water ice. The ground beneath the boulders is a little softer, suggesting that it might be more like a frozen crust of sand with bits of liquid below, presumably liquid methane.

Neither *Huygens* nor *Cassini* have revealed anything that looks like evidence of life, but Titan remains one of the leading candidates for life in our solar system. In addition to the possibility of cold life that uses liquid methane rather than liquid water, Titan even offers a small chance of water-based life in at least two ways. First, the evidence of icy volcanism suggests there could be "hot" springs where the temperature rises to slightly above the freezing point, allowing liquid water to exist. In addition, Titan's interior contains ices to great depths, so there is almost certainly a subsurface layer where the ices are melted. Titan might even have a subsurface ocean like that on the Galilean moons, though it would probably lie much deeper beneath the surface and would more likely be a cold, ammonia-water mixture than a warmer ocean of just plain water. The organic molecules produced naturally in Titan's atmosphere could provide nutrients for life, as might chemical reactions that could occur between liquid and rock near sources of icy volcanism. Sadly, it will probably be quite a while before budgets allow us to land more sophisticated robots on Titan, but over the longer term, I suspect we humans will continue to be drawn by the Sirens of Titan.

THE FOUNTAINS OF ENCELADUS

No other moon of Saturn is even close in size to Titan or to any of Jupiter's Galilean moons. Rhea, the next largest of Saturn's moons after Titan, is less than 1/3 of Titan's size in diameter and less than ½ the diameter of Europa (smallest of the Galilean moons). We expect that such relatively small moons should retain far less radioactive heat than larger moons, and any tidal heating also should be at least slightly less effective. Nevertheless, several of Saturn's moons show evidence of past geological activity, and scientists got a huge surprise in 2005, when *Cassini* observed plumes of icy spray shooting out into space from the surface of the moon named Enceladus (color plate 7).

Enceladus is barely 300 miles in diameter, so you could almost set it down within the borders of Colorado. Prior to the *Cassini* observations, most scientists would have said that this is simply too small to allow any possibility of life, because all such moons should be frozen solid. But Enceladus has proven otherwise. *Cassini* photos show a moon with few craters, indicating recent repaving by geological activity. The most startling regions are the bluish "tiger stripes" (see color plate 7) near the moon's south pole. The tiger stripes are measurably warmer than the surrounding terrain, and close-up examination suggests that they are cracks or grooves through which material can well up from below. These regions appear to be covered by "fresh" ice—ice that has solidified on the surface within no more than the past few thousand years, and possibly within just the past few decades or less. Moreover, images taken by *Cassini* as it looked at Enceladus backlit by the Sun show that fountains of ice particles and water vapor spray out from the tiger stripe regions. This is clear evidence of icy volcanism on Enceladus. Spectral analysis of the tiger stripe regions also shows some simple organic molecules.

The mechanism driving the geyser-like spray is still being debated, but it is almost certainly related to tidal heating of some type. Moreover, while ice could sublimate directly to water vapor in making the spray, models suggest that Enceladus must have a subsurface liquid layer deep under its surface. The liquid is probably an ammonia-water mixture, but it's conceivable that it could be more pure water. The astonishing conclusion: This small moon could have subsurface habitable zones. Moreover, the fact that we have been so surprised by Enceladus should tell us that other surprises are likely to await us. Our basic ideas about where to look for life are probably still valid as starting points, but it might be wise to keep an open mind about other places as well.

LIFE BEYOND SATURN

As we've moved outward on our biological tour, we've also moved down the scale of likely habitability. Mars seems so likely to be habitable that I'll be surprised if it does *not* turn out to have life. Europa seems reasonably likely to be habitable, with its neighbors Ganymede and Callisto somewhat less so. Titan and Enceladus also offer possibilities for life, though with their cold temperatures they seem to stretch the limits. Nevertheless, our surprising findings, especially on Enceladus, make us ask whether there could be still other habitable places in our solar system.

After the Galilean moons and Titan, the next largest moon in our solar system is Neptune's moon Triton. Triton is unusual for a large moon in that it orbits Neptune "backward"—that is, in a direction opposite to the direction that Neptune rotates. This type of backward orbit is a telltale sign of an object that once orbited the Sun independently and later was captured into planetary orbit. It's not easy for a planet to capture a moon of any size, let alone a moon as big as Triton, and scientists are struggling to explain how it might have happened (though some recent ideas look plausible). Nevertheless, its unusual orbit tells us that Triton once orbited the Sun as a "planet" (or dwarf planet) about 20 percent larger than Pluto.

Triton has been photographed close-up only once, by *Voyager 2* in 1989. Its icy surface is smooth in some places and in others is crinkled into patterns resembling the skin of a cantaloupe (figure 7.3). There are few impact craters, so Triton has apparently had geological activity within at least the past 100 million years or so. What drives this geological activity? Triton must have some internal heat, possibly including heat left over from tidal heating that would have occurred in the past, as it was being captured into its present orbit. Some researchers think this heat may be sufficient to cause ice volcanoes occasionally to erupt from a liquid ocean beneath the surface. The liquid would be much, much colder than ordinary liquid water, and probably would consist of water mixed with ammonia, methane, or other melted ices. Nevertheless, if it's liquid, we have at least a slim possibility of life.

Aside from Triton, the rest of the moons of Uranus and Neptune seem too small to have liquid regions or any life—but, of course, that's what we thought about Enceladus before we went there. Some day, when we finally get around to sending spacecraft back to Uranus and Neptune, we'll find out whether life could exist around these large but distant planets.

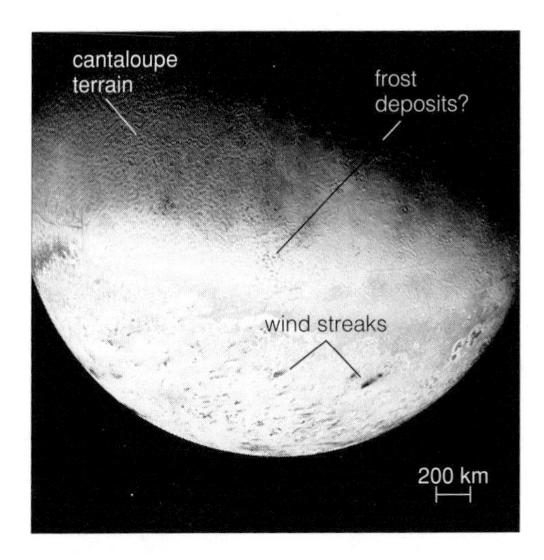

Figure 7.3. The southern hemisphere of Neptune's moon Triton, photographed by *Voyager 2* in 1989. (Courtesy NASA.)

BEYOND UFOS

We do not yet know of life anywhere besides Earth. But let's just count the possibilities we've found in our own solar system. We've found at least some evidence that liquid water exists on Mars, Europa, Ganymede, Callisto, Titan, and Enceladus. That's six worlds right there, and we can add Triton if we allow for colder liquids. And there could be other possibilities that we have not yet identified.

Notice that, except for Mars, all these potentially habitable worlds are *moons* orbiting large planets. The lesson should be clear: If similar moons are common around the planets of other stars—and we have every reason to think they should be—such moons might be the most common homes to life in the universe.

The possibilities here in our solar system also raise the intriguing prospect that we might soon be able to conduct true comparative biology. While life on Mars could potentially be transplanted Earth life (if it migrated there on meteorites), it's difficult to see how any of these other worlds could have anything besides indigenous life. If life exists in any of these places, it will almost certainly give us the opportunity to study an entirely different biochemistry from that which we find here on Earth.

The ongoing exploration of the solar system may therefore someday yield a research gold mine for biologists and biochemists, not to mention planetary scientists. But the sociologists, anthropologists, and political scientists may feel a bit left out in the cold, because we won't be finding life in our solar system that we can talk to. For that, we must look to the stars.

8

LIFE AMONG THE STARS

> How vast those Orbs must be, and how inconsiderable this Earth, the Theatre upon which all our mighty Designs, all our Navigations, and all our Wars are transacted, is when compared to them. A very fit consideration, and matter of Reflection, for those Kings and Princes who sacrifice the Lives of so many People, only to flatter their Ambition in being Masters of some pitiful corner of this small Spot.
>
> —*Christiaan Huygens, c. 1690*

In 1999, I had the good fortune of spending a few months living in Holland, with my wife and then one-year-old son. We lived in a small town called Aalsmeer, just outside Amsterdam and even closer to the Schiphol airport. We chose the town because my wife's employer had an office there, which was why we had come to Holland in the first place, but we soon learned that it is most famous as the site of the world's largest flower market. As a fairly small town, Aalsmeer gave us a sense of the way most Dutch people really live, which is not quite according to the free-living reputation of Amsterdam. It was a great place to be with a young child, and its proximity to the Dutch highways made it a great starting point for driving trips throughout the region. For Americans, Europe seems remarkably small. In less time than it takes us to drive from our home in Boulder to the Colorado border, we could pass through three, four, or even five countries in Europe.

Today, crossing between Holland and Belgium or France and Germany is little different than crossing the Colorado border into New Mexico or Utah. If you're lucky, you'll see a sign letting you know when you've actually passed from one country to the other. In the town of Echternach, in Luxembourg, one day I pushed my son in his stroller across a short footbridge to

Germany, where he could have had a glimpse of his fifth country if not for the fact that he was sound asleep. And as I looked at him sleeping peaceably, it hit me: When his grandparents were children, people on opposite sides of this bridge were killing each other. Just decades ago, every one of those border crossings that we make so easily today would have taken us from one warring country to another, putting our lives at risk. Going back just a few centuries further—easy to do in Europe, as you walk among the ruins of fortified castles—you might have been in grave danger just by climbing the nearest hillside. The modern world has a lot of problems yet to solve, but as a species we've made some remarkable progress. Historic enemies, like the French and the Germans or the Sienese and the Florentines, may still sometimes say nasty things about each other, but they no longer seem inclined to go to war over every transgression, real or imagined.

This welcome change has probably come about for a great many intertwined reasons. Economists will point to the importance of growing trade, historians will tell you of the role of culture and treaties, and political scientists can discuss the importance of democracy and of expanding views of human rights. To this list, I'll add something that is surely at least as important, though rarely mentioned: science, and especially the new perspective it has brought us on our place in the universe. After all, as described so eloquently in the quotation at the beginning of this chapter from the Dutch scientist Christiaan Huygens—whose home-built instruments and telescopes are well worth seeing in the Museum Boerhaave in Leiden (less than an hour's drive from Aalsmeer)—a little perspective on our place in the universe makes any form of geographic, ethnic, or religious hatred seem just plain ridiculous.

Of course, plenty of educated people have participated in the atrocities of the past and present, even though they presumably were taught that Earth is a planet orbiting the Sun. But, personally, I suspect that for these people, the lesson never set in, and they suffer from what I call "center of the universe syndrome." They've never grown up, and like young children they still imagine themselves to be the center of everything. Show me a Saddam Hussein or a Kim Jong-Il or any other petty dictator, and I'll show you a person who suffers from center of the universe syndrome. How else to explain their belief in their own self-importance? I tie the syndrome to crime, as well. Many of the inner city youths who get caught up in gangs live in a universe that, for them, consists of little more than a few square miles of urban landscape; many of them have never been far enough from city lights to truly see the stars. Indeed, I have friends who've taken such kids on camping trips, and my friends report incredible experiences, such as gang-hardened thugs expressing fear—and, sometimes, an almost immediate change in their hearts—as they finally see the stars and realize how much more there is to this world than they had ever before imagined.

Getting over center of the universe syndrome is not easy. Some people seem to suffer from it no matter how much they have learned. Incredibly to me, there are even scientists who suffer from it: Apparently, rather than getting a sense of awe and humility from their studies of nature, they become enamored with their own research accomplishments, sometimes to the point of attacking others who may disagree with their conclusions, or denigrating students who they wrongly believe to be intellectually inferior. The syndrome is difficult to beat because we're all born with it, and it's just human nature to want to keep thinking of oneself as somehow special, better, or more important than others. But most of us eventually learn that other people have the same thoughts and feelings as we do, and therefore learn empathy and, with it, the importance of treating our fellow humans with kindness and respect. In other words, most of us eventually grow up.

And that is where the topic of this chapter comes in: The human race is growing up, too. We once thought that this world was all there was, so perhaps it was natural to fight over every piece of it. With the Copernican revolution, we began to realize that there might be more, but how much more remained unknown. Perhaps the first true glimpse of infinity came with Christiaan Huygens, who wrote the passage that I quoted not long after he became the first person to make reasonably accurate estimates of distances to the stars. He was therefore the first to understand, at least with any scientific certainty, that other stars might truly be other suns, orbited by planets and, perhaps, harboring their own life. But this last "perhaps" tells us that we haven't yet matured beyond our adolescence as a species: While we've reached the point where we understand that our world is small and precious *to us*, we don't yet know if we are members of a larger community. In my opinion, we can't finish growing up unless we continue the effort to find out, one way or the other.

DISTANT SUNS

If we're going to search for life among the stars, we need to know where to look. The first step in this process is to decide which stars are good candidates for having habitable planets in orbit around them—that is, to know which stars would make good "suns." To do this, we must understand a little bit about the nature of stars.

In ancient times, almost any light in the sky was considered to be a star, and in some cases we still use this historical language. For example, we often refer to meteors as "shooting stars," even though they really are just bits of interplanetary dust entering our atmosphere. Asteroids got their name, which means "star-like," because that is how they appear when first seen through a telescope, even though they are actually chunks of rock in our own solar system. Our modern definition of a star is a large ball of gas that, like our Sun, generates energy by nuclear fusion in its hot central core.

Stars are not living organisms, but they nonetheless go through life cycles. All stars are born from the gravitational collapse of large clouds of interstellar dust and gas, just as our Sun was born some 4½ billion years ago. As gravity compresses a star, its insides get hotter and denser. A star is "born" when its core becomes hot enough to start the fusion reactions that power it throughout its life. The star then shines until it ultimately runs out of fuel for fusion, at which point it dies, scattering some of its remains back into space where they can be recycled into later generations of stars, while leaving the rest behind as a stellar corpse that may either be a white dwarf, a neutron star, or a black hole.

For astronomers, all aspects of stellar lives are fascinating, and perhaps none more so than the strange stellar corpses—especially the black holes—that test the limits of our understanding of physics. But for our purposes in searching for organic life, the only part that really matters is the time during which the star shines with energy from nuclear fusion. Although this might seem obvious given the importance of sunlight to life on Earth, it's actually a bit subtle.

Remember that, as we discussed in the prior chapter, several of the potentially habitable worlds in our solar system—including Europa, Ganymede, Callisto, and Enceladus—have at least some chance of harboring life that does not depend on sunlight at all. Instead, this life would live off energy generated internally, by tidal heating or radioactive decay. These types of worlds could in principle orbit almost any massive object, star or not. For example, when the Sun dies and leaves a white dwarf as a remnant some 5 billion years from now, it's quite possible that Jupiter will continue to orbit the white dwarf, with its moons intact. In that case, Europa will still be tidally heated through the orbital resonance it shares with Io and Ganymede, and life in the Europan ocean, if it exists, may not even notice that the Sun has gone out.

Similarly, Jupiter-like planets with tidally heated moons could exist around objects too small to shine as stars. Calculations show that in order for gravity to compress an object to the point where its center reaches the temperatures needed for nuclear fusion, the object must have a mass at least 8 percent that of the Sun (which is equivalent to about 80 times the mass of Jupiter). If it is smaller than that, it will never become quite hot enough to sustain fusion. We now know that such "substellar" objects are common in the universe; we call these objects *brown dwarfs,* and we have already found that some of them are indeed orbited by large planets. Our technology is not yet capable of determining whether any of these planets have moons, but there's no reason to think they wouldn't. And if they have multiple moons, these moons are likely to share orbital resonances that could lead to tidal heating like that on Io, Europa, and Ganymede. Life could well prove to be common on such worlds.

However, while all this subsurface life would surely interest biologists, it probably has no relevance to the search for other intelligent beings. As I've argued in prior chapters, intelligent life probably can evolve only on a planet with *surface* liquid water and abundant sunlight. So when we restrict the search for habitable worlds to those with habitable *surfaces,* by necessity we are looking for worlds that orbit a shining star, not a brown dwarf or a stellar corpse.

This might not seem at all restrictive at first, because one look at the night sky tells us that there are lots of shining stars. But stars are not all the same. Stars vary significantly in their masses, temperatures, and luminosities, and in the dominant type of light they emit (such as visible light versus ultraviolet light). These differences could have significance to the potential for life.

The most important factor is mass. Stars range in mass from the smallest true stars that are only 8 percent the mass of the Sun up to monsters that can be up to about 150 times the mass of the Sun. For reasons I won't go into here, mass turns out to control all of a star's other properties, including its temperature and luminosity. And of direct relevance to the search for life, stellar mass determines stellar lifetime: *The more massive the star, the shorter its life.* This might seem counterintuitive, since more massive stars obviously have more fuel to burn through nuclear fusion. However, the stronger gravity of more massive stars also makes them much hotter inside, and the higher temperature means they fuse their hydrogen into helium at a far greater rate than do smaller stars. The result is that massive stars live fast and die young. That is, even though they have more fuel available, they burn through it so fast that they run out of it in a relatively short time.

The most massive stars live only a few million years or less—which immediately rules them out as potential "suns" for habitable worlds. A few million years isn't even enough time for Earth-like planets to form in the first place, let alone to develop life. A star would have to live at least a hundred million years or so just to give time for planets to form and for life to get started. Even then, the example of Earth suggests that a star would have to live at least a couple of *billion* years to give life a moderate chance of getting much past the microbial stage. This constraint rules out all stars more massive than about twice the Sun's mass.

One question that may occur to you at this point is how many stars this mass constraint has ruled out, and it gives me a good opportunity to introduce the idea of what scientists call a *selection effect.* If you consider the stars we can see with the naked eye, ruling out those with more than twice the Sun's mass means ruling out almost all of them. However, this turns out to be completely *un*representative of stars as a whole. Remember, more massive stars burn through their fuel at a much greater rate than the Sun, which also means they are much brighter and hence much easier to see from great distances. For example, if you took the Sun and placed it 50 light-years away, it would barely be visible to the eye even on the clearest and darkest nights. In contrast, a star with 25 times the mass of the Sun may shine brightly in our sky even if it is hundreds of light-years away. In fact, most lower-mass stars are dim enough that we may not be able to see them even when they are quite nearby; the small star Proxima Centauri (one of the three stars of the Alpha Centauri system) is so dim that it is invisible to the naked eye, despite being the nearest of all stars besides the Sun. So this is what we mean by a selection effect: A star is far more likely to be "selected" as a member of the group we can see by eye if it is unusually massive (and therefore unusually luminous) than if it is more ordinary. Only as our telescopes have improved, enabling us to see the dimmer stars, have we been able to make a more realistic inventory of stellar masses. The results show that high-mass stars are comparatively rare. In ruling out stars with more than twice the mass of the Sun, we've ruled out no more than about 1 percent of all stars.

Stars similar to the Sun in mass clearly qualify as potential suns, since they would have lifetimes of many billions of years and habitable zones as wide as that of the Sun, offering ample room for the existence of Earth-like planets. Depending on how wide a range you choose for "similar" to the Sun, this group may represent between 5 percent and 10 percent of all stars. The rest, which means 90 percent or more of all stars, are in the low-

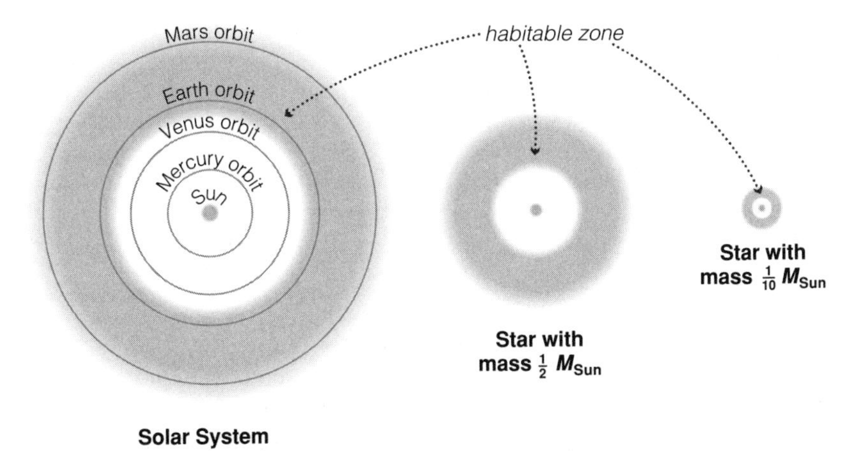

Figure 8.1. The approximate locations and sizes, to scale, of habitable zones around the Sun and two stars of lower mass (and hence lower luminosity). (Illustration courtesy of Addison Wesley, an imprint of Pearson Education)

mass group. What can we say about the possibility of habitable planets around these most common of stars?

On the one hand, they clearly have lifetime going for them. Any star less massive than our Sun will live longer than our Sun's 10 billion years, and the very low-mass stars may live hundreds of billions of years—many times the current age of the universe—before ultimately burning out. Low-mass stars offer plenty of time for evolution. On the other hand, because their total light output is so much smaller than that of the Sun, these low-mass stars must have much smaller habitable zones (figure 8.1). Recall that the habitable zone represents the range of distances from a star at which the strength of sunlight would be enough to warm a planet so it could have liquid water on its surface, but not so great as to cause a runaway greenhouse effect. Just as a living room fireplace offers a much smaller and closer-in range of distances than a bonfire at which you can sit and appreciate the warmth, a dimmer star must have a much closer-in and narrower habitable zone than a star like the Sun.

If you run through the calculations and assume that planets are equally likely to form at any distance around a star, you'll find that the chance of a planet forming in the habitable zone around a "typical" small star (with about 10 percent the mass of the Sun) is only about 1/20 that of forming in the habitable zone around a Sun-like star. This means that, all other things being equal, these small stars are only about 1/20 as likely to have habitable planets as Sun-like stars. However, what these stars lack in brightness they make up for in numbers. Because these stars are also 10–20 times as common as Sun-like stars, we conclude that, roughly speaking, there could be just as many habitable planets around small stars as around larger ones.

Until recently, I might have needed to temper this somewhat optimistic view of the possibility of life on planets around small stars for two reasons. First, the habitable zones of these stars are so close in that planets in these zones would almost certainly be locked into synchronous rotation, with one side perpetually facing the star in the same way that one side of the Moon stays facing toward Earth. This seemed to imply that one side of such a planet would get way too hot while the other side remained perpetually dark and cold. Second, small stars tend to produce more frequent and energetic flares than stars like our Sun, which created a worry that life on a close-in planet would be fried by the intense ultraviolet light and X rays produced during flares. However, most scientists no longer think that either of these concerns poses a significant problem for life. Recent research suggests that even a modestly thick carbon dioxide atmosphere would circulate heat from the bright to the dark side of a synchronously rotating planet, keeping temperatures relatively uniform and potentially allowing liquid water to exist over much of the planet's surface. As for the flares, a planetary atmosphere could also provide protection against this radiation, and in any event the discovery of radiation-resistant organisms such as *D. radiodurans* makes life seem a lot more resilient than we used to think. Moreover, as long as the radiation didn't kill the life, it might spur a higher mutation rate, perhaps making evolution proceed faster on such worlds than it did on Earth.

So far, we've discussed only single stars, finding that all but the rare massive ones could at least potentially be orbited by habitable planets. But another consideration is the fact that most stars are not single, like our Sun; instead, somewhat over half of all stars are members of multiple star systems, in which two or more stars orbit each other. In some cases, the competing gravitational pulls in such systems could make it impossible for a planet to have a stable orbit in a habitable zone. For example, if our own Sun had a companion star orbiting at, say, the distance of Mars, Earth would have been thrown off course long ago. These days, however, scientists can calculate orbital properties in multiple star systems without too much difficulty, and the results suggest that multiplicity probably has little overall effect on the prospects for habitable worlds. Many star systems are close binaries, in which two stars orbit each other at much closer distance than Mercury orbits the Sun. Planets in such systems could have stable orbits around the two stars together. Beings on such worlds would simply see two stars rising and setting in tandem in their sky; you can see such a scene in the original *Star Wars,* in which the planet Tatooine has twin suns. In other cases, where the two or more stars are widely separated, a planet could orbit just one of the stars, feeling little gravitational effect from the others. Overall, there seems no reason to think that the tendency of stars to be paired up or living polygamously will substantially reduce the chances of finding life beyond Earth.

In wrapping up this discussion of distant suns, I should acknowledge that these conclusions are based on the assumption that having a habitable planet only requires having a habitable zone in which stable orbits are possible. However, a few scientists have proposed that it might be a lot more difficult than that. I'll discuss this "rare Earth hypothesis" shortly, but first let's continue with the assumption that plenty of stars could make good suns.

FINDING PLANETS

Based on the assumptions I've used, almost all stars should be capable of having habitable planets. If the Sun's one habitable planet (Earth) is typical for Sun-like stars, then we'd expect that at least 5–10 percent of the stars in our galaxy (and other galaxies) would have a habitable planet, and perhaps 1 out of 10 or 20 of the more common but smaller stars would also have a habitable planet. The implied numbers are staggering. You may recall that, earlier in the book, I said that our galaxy has "at least" 100 billion stars. A more careful accounting suggests that the actual number is somewhere between about 500 billion and 1 trillion stars. Even if we take the lower value and assume that only 5 percent of stars have a habitable planet, we are still talking 25 billion habitable worlds. It takes only a slightly less conservative view to raise the estimate to 100 billion habitable worlds. That means we can repeat the counting exercise that I offered in chapter 1, and we'd find that just *counting* the habitable planets in our galaxy might take some 3,000 years, let alone detecting each of them and then studying them in enough depth to determine whether they are home to life.

Nevertheless, not much more than a decade ago, we did not know with certainty whether *any* planets existed beyond our own solar system, habitable or not. The problem was, and still is, that seeing planets around other stars is extraordinarily difficult. If you think back to the Voyage scale model of the solar system that we discussed in chapter 3, you'll realize that seeing an Earth-like planet orbiting even the nearest stars would be somewhat like looking from San Francisco for a pinhead orbiting just 15 meters from a bright grapefruit in Washington, D.C. Seeing Jupiter would be only marginally less difficult, as it would mean looking for a marble located about a football field away from the bright grapefruit. The scale alone would make the task quite challenging, but it is further complicated by the fact that a Sun-like star would be a *billion* times as bright as the light reflected from any of its planets. Because even the best telescopes blur the light from stars at least a little, the glare of scattered starlight would tend to overwhelm the dim points of planetary light.

If you think about these challenges, you might easily come to the conclusion that planet detection would be nearly impossible. But we've done it. The first clear-cut discovery came in 1995, and since that time discoveries of extrasolar planets have poured in at an astonishing rate. To give you a personal viewpoint on it, let me tell you a quick anecdote about my astronomy textbook. When we wrote the first edition, which came out in 1998, we included a nice little chart with a row for each of the 13 extrasolar planets that had been discovered by that time, visually showing the planet's mass and distance from its star. By the second edition, the chart had expanded to show 55 planets, and by the third edition, which came out in 2003, it showed 77 planets. The rows were starting to get awfully small and close together in order to still fit on a page. For the fourth edition, the number had grown past 150, and it has surpassed 250 as we work on the fifth edition. We therefore have had to abandon the chart altogether and develop new ways of showing the growing statistical data about extrasolar planets. Indeed, our knowledge of other planetary systems has reached the point where we stopped just devoting a few pages and a chart to them, and instead wrote a brand-new chapter about them. I know that people sometimes complain that publishers revise textbooks just so they can sell new books, but in a field like astronomy (or astrobiology), we authors can barely keep up; indeed, we've many times been tempted to call our colleagues and ask them to please stop discovering new things for a few weeks, just so we can write *something* that will be momentarily up-to-date. In these fields, at least, it would be a grave disservice to college students to ask them to use a textbook more than two or three years old.

The rapid rate at which we are discovering planets around other stars leaves no room for doubt that other planetary systems exist and are common. And yet, among these couple of hundred recently discovered planets, not a single one is an Earth-size planet orbiting in its star's habitable zone. Does that mean that we've been wrong in inferring that such planets ought to be common, and that in reality they are extremely rare? It's possible, but it seems much more likely that we're dealing with another selection effect. Planets as small as Earth are far more difficult to detect than larger planets, so their apparent absence is probably just a reflection of the fact that our technology is not yet up to the task. To understand why, and how we hope to change that soon, it's worth spending a little time talking about the techniques used to find planets around other stars.

If you strip away the details, there are just two basic ways of detecting extrasolar planets: directly and indirectly. Direct detection means an actual photograph, and in principle it would be the method of choice. However, because of the problems of scale and glare, our current telescopes are really not yet capable of taking such direct photographs. The only exceptions as of the time I write have been cases of planets orbiting brown dwarfs, which pose far less of a glare problem, since they are not truly stars. Other than that, direct detection remains a technology of the future. That is why nearly all known extrasolar planets have been discovered by some type of *indirect* technique.

Indirect detection essentially means a way of discovering a planet by observing its effect on its star rather than by seeing the planet itself. There are at least two different ways that a planet can cause a noticeable change in its star. One is by causing its star to "wobble" as a result of the planet's gravitational influence. Another is by causing its star to sometimes get a bit dimmer as the planet passes in front of it and blocks some of the star's light.[1] Let's start with the wobble idea, since it is behind the vast majority of extrasolar planet discoveries to date.

Although we usually think of a star as remaining still while planets orbit around it, that is only approximately correct. In reality, all the objects in a star system, including the star itself, orbit the system's *center of mass*. To understand this concept, think of a waiter carrying a tray of drinks. To carry the tray, she places her hand under the spot at which it balances—its center of mass. If the tray has a heavy glass of water off to one side, she will place her hand a little to that side of the tray's center. Because the Sun is far more massive than all the planets combined, the center of mass of our solar

[1] For the sake of completeness, I should note that there are a few other indirect techniques that astronomers have come up with, including a technique called *gravitational lensing* that has been used to discover a planet only about five times as massive as Earth.

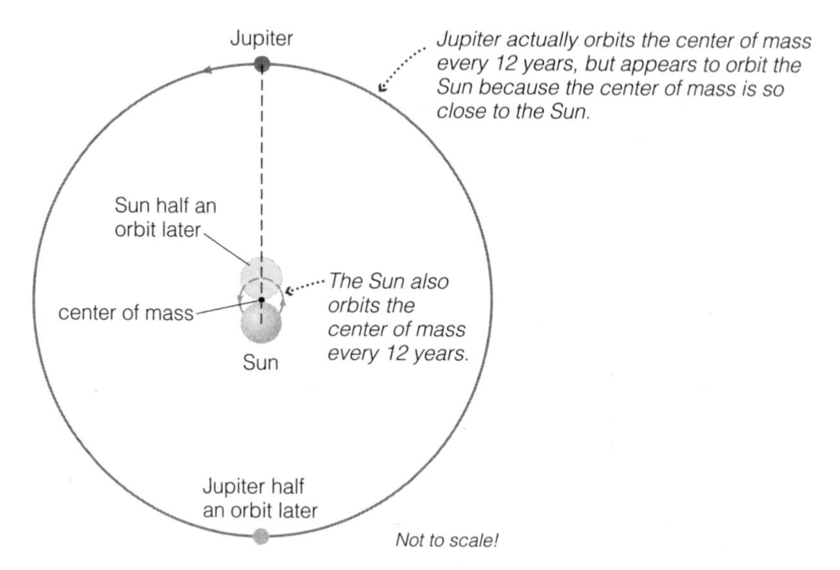

Figure 8.2. This diagram shows how both the Sun and Jupiter actually orbit around their mutual center of mass, which lies very close to the Sun. The diagram is not to scale; the sizes of the Sun and its orbit are exaggerated about 100 times compared to the size shown for Jupiter's orbit. (Illustration courtesy of Addison Wesley, an imprint of Pearson Education)

system lies very close to the Sun—but not exactly at the Sun's center. We can see how this fact allows us to discover extrasolar planets by imagining the viewpoint of extraterrestrial astronomers observing our own solar system from afar.

Let's start by considering only the influence of Jupiter, which exerts a stronger gravitational tug on the Sun than the rest of the planets combined. The center of mass between the Sun and Jupiter lies just outside the Sun's visible surface (figure 8.2). In other words, what we usually think of as Jupiter's twelve-year orbit around the Sun is really a twelve-year orbit around their mutual center of mass; we generally don't notice this fact because the center of mass is so close to the Sun itself. In addition, because the Sun and Jupiter are always on opposite sides of the center of mass (otherwise it wouldn't be a "center"), the Sun must orbit this point with the same twelve-year period as Jupiter. The Sun's orbit traces out only a very small circle (or ellipse) with each twelve-year period, because the Sun's average orbital distance is barely larger than its own radius. Nevertheless, with sufficiently precise measurements, extraterrestrial astronomers could detect this orbital movement of the Sun. They could thereby deduce the existence of Jupiter, even without having observed Jupiter itself. They could even determine Jupiter's mass from the Sun's orbital characteristics as it goes around the center of mass, because for any particular separation between the Sun and a planet, a more massive planet would pull the Sun around its small orbit at a faster orbital speed.

Now let's add in the effects of Saturn, which would have the second most noticeable gravitational influence on the Sun. Saturn takes 29.5 years to orbit the Sun, so by itself it would cause the Sun to orbit their mutual center of mass every 29.5 years. However, because Saturn's influence is secondary to that of Jupiter, this 29.5-year period appears as an added complication on top of the Sun's twelve-year orbit around its center of mass with Jupiter. In other words, every twelve years the Sun would return to *nearly* the same orbital position around its mutual center of mass with Jupiter, but the precise point of return would make its own even tinier orbital circle (or ellipse) with Saturn's 29.5-year period. By measuring this motion carefully from afar for a few decades, an extraterrestrial astronomer could deduce the existence and masses of both Jupiter and Saturn.

The other planets also exert gravitational tugs on the Sun, which create further complications to the Sun's orbital motion around the solar system's center of mass. These extra effects would become increasingly difficult to measure in practice, but in principle they would allow an extraterrestrial astronomer to discover all the planets in our solar system. If we turn this idea around, you'll realize that it means we can search for planets in other star systems by carefully watching for the tiny orbital motions of a star around the center of mass of its star system.

You might guess that the easiest way to notice this stellar motion would be to make very precise measurements of stellar positions in the sky, so that we could in essence watch as a star traces its small circles around the center of mass. However, with current technology, this stellar motion is measurable only if the star traces a fairly large orbit around the center of mass, which means it works best for finding large planets that are far from their stars. To understand why, consider what would happen if Jupiter moved farther from the Sun. Moving Jupiter outward from the Sun would cause their mutual center of mass to move farther from the Sun, making the Sun's orbital motion larger and easier to detect. However, at the same time, moving Jupiter outward would also cause its orbital period to get longer, and the Sun's orbital period around the center of mass would increase along with it. The practical effect is that it would take much longer to detect Jupiter's influence. In its actual location, Jupiter and the Sun orbit their mutual center of mass every twelve years, so it takes only twelve years to observe a full orbit of the Sun around this point. But if Jupiter moved to, say, the 165-year orbit of Neptune, it would take 165 years to observe a complete orbit. While we might not need to watch for quite that long to realize we are seeing orbital motion, we'd probably have to watch at least half an orbit—some 80 years in this case—to be confident of what we are seeing. As a result of this practical limitation, this technique (known as the *astrometric* technique) has to date been used for only a handful of extrasolar planet discoveries.

Fortunately, there's another, easier way of measuring a star's motion around the center of mass, and it works by taking advantage of something called the *Doppler effect*. You're probably familiar with the Doppler effect for sound: It's responsible for the change in pitch—the sort of "weeeeeeee-ooooooooooh" sound—that you hear from a train whistle as it passes by you on the train tracks. (You can also notice the Doppler effect with emergency sirens or even just with the "buzz" of a fast car as it goes past you.) The Doppler effect occurs because of the way sound waves are affected by motion. When a train is moving toward you, its sound waves are effectively bunched up as they move in your direction, causing the higher pitch. When the train is moving away from you, the sound waves are stretched out behind it, making the pitch lower. The Doppler effect causes similar shifts in the wavelengths of light. As a result, when a star is in the portion of its orbit where it is coming toward us, its light is shifted to slightly higher frequencies (a *blueshift*); when it is on the other side of its orbit, moving away from us, its light is shifted to slightly lower frequencies (a *redshift*).[2] It is observations of these types of slight frequency shifts that have led to the vast majority of the planet detections so far.

And here's where the selection effect comes in: Current technology is remarkable in being able to measure Doppler shifts in stellar spectra for

[2] Note that we can observe Doppler shifts only if the star has at least some of its orbital motion directed toward and away from us; that is, we cannot observe any Doppler shift for a star and planet that have their mutual orbits oriented face-on toward Earth, since such orbits would just go in small circles on the sky, with no component of motion coming at or moving away from us. In addition, because Doppler measurements tell us only the portion of the star's total orbital speed directed toward and away from us, they tend to underestimate the true orbital speed. A consequence of this fact is that the Doppler technique gives us only lower limits on planetary masses, rather than more precise estimates, except in rare cases for which we have some way of knowing the orbital inclination.

stars orbiting the center of mass at speeds as slow as about 1 meter per second—walking speed—but even this speed occurs only under the influence of a fairly massive planet orbiting fairly close to its star. That is, the Doppler technique preferentially "selects" for large and close-in planets, and is not yet sensitive enough to reveal the existence of planets as small as Earth. Moreover, because the orbit of a small planet like Earth would most likely appear only as a tiny complication to the orbital influence of larger planets around the same star, it's likely to be quite a while before our technology reaches the point at which we could use this technique to find Earth-like planets. The absence of Earth-size planets known to date is probably only a reflection of the limitations of the Doppler technique.

THE SEARCH FOR EARTH-SIZE PLANETS

So how might we learn whether Earth-size planets exist? Earlier, I noted that in addition to causing a star to wobble through its gravitational influence, there is another way in which a planet might cause a noticeable effect on its star: The planet might block some of the star's light as it passes in front of it, in which case the dip in the star's brightness would offer indirect evidence of the planet's existence. Observations of such dips are probably the way we will make the first discoveries of Earth-size planets around other stars.

Before we get into details, note that a planet can pass in front of its star as seen from Earth only if its orbit happens to be oriented edge-on in our sky. In essence, the orbital geometry must be perfect for an eclipse, just as the Moon can blot out the Sun in the sky only on those relatively rare occasions when it passes directly between us and the Sun. Because a planet is much smaller than a star and can never fully block its star's light, a better analogy is to the even rarer occasions on which Mercury or Venus pass across the face of the Sun as seen from Earth. We call such events *transits;* for example, Venus transited across the face of the Sun on June 8, 2004, and will do so again on June 6, 2012—and then there will be no more transits of Venus for another 105 years.

The somewhat odd pattern in the timing of the transits of Venus comes about because we are also orbiting the Sun, and our orbit is inclined to Venus's orbit. (For similar reasons, the timing of Mercury transits is also complex.) But this type of complicated pattern won't arise when we look at planets orbiting other stars, since their great distances mean we are essentially stationary in comparison. Instead, for any extrasolar planet that has its orbit oriented edge-on in the sky, we should be able to see a transit once every orbit. In other words, if we keep watching, the transits will repeat at precise intervals. This is important, because if we noticed just a single dip in brightness for a star, or even many randomly timed dips, we'd have no way of knowing whether the dip was caused by a planet or because the star, for some other reason, actually got temporarily dimmer. However, if we see dips of the same amount repeating, say, every four months, we know we have caught a planet that orbits the star every four months. Moreover, the size of the dips tells us the size of the planet. For example, if the dips reduce the star's brightness by 1 percent, then we know the planet is large enough to be blocking out 1 percent of the face of the star.[3]

If you think about these ideas, you'll realize that there are three limitations on the transit method for planet detection. First, the planet must have its orbit oriented almost precisely edge-on, or else transits will never occur. Second, we must have instruments sensitive enough to measure very small changes in stellar brightness. Third, we must observe long enough to see the brightness changes repeat, so we can be confident they are due to a transiting planet rather than being intrinsic brightness variations of the star. These limitations are significant but not insurmountable.

The first limitation might seem fairly severe. After all, there is no reason for the orbits of planets around other stars to have any particular orientation relative to us, so it's just random chance that determines when a star happens to have planets with edge-on orbits as seen from Earth. For a planet the size of Earth, random chance implies that only one in several hundred planets would have an orbit with the right orientation. The chance rises a bit for larger planets, but it's still an inescapable fact that the transit technique has no chance at all of finding more than 99 percent of the planets that exist. Nevertheless, as in much of astronomy, numbers work in our favor. If we search for transits around, say, 100,000 stars, a 1 percent hit rate will yield 1,000 planets.

In fact, because scientists have indeed monitored a lot of stars, we have already discovered a few extrasolar planets with the transit technique. However, these planets are all fairly large so far, because of the limitation on measuring brightness dips. A planet the size of Jupiter will typically block 1 percent of its star's light, which is measurable with ground-based telescopes. But a planet the size of Earth would block less than 0.01 percent

[3] We have independent ways of knowing stellar radii, so once we know the percentage of the star's face that is blocked by the planet, it's straightforward to calculate the planet's precise radius.

of its star's light, which is difficult to measure with telescopes on the ground, because effects due to Earth's atmosphere can overwhelm this small change. If we want to observe Earth-size planets with transits, what we really need is an extensive star monitoring program carried out with a telescope in space. In other words, we need the mission that NASA is calling *Kepler*.

The *Kepler* mission, currently slated for launch in 2009, will spend four years carefully monitoring the brightnesses of some 100,000 stars. It will be capable of bagging transiting planets even smaller than Earth. If solar systems like our own are fairly typical, then over its four years of observations, Kepler should detect many hundreds of large planets, and perhaps 50 or so Earth-size planets.[4] If it really does detect this many or more Earth-size planets, we'll know that Earth-size planets are indeed quite common in the universe. So I'll ask you to return to the quotation that opened chapter 1, in which Saint Albertus Magnus asked: "Do there exist many worlds, or is there but a single world?" Presuming he meant worlds like ours, it now seems likely that we'll know the answer to this question in less than a decade.

ARE EARTH-LIKE PLANETS RARE OR COMMON?

Kepler should tell us whether Earth-*size* planets are as common as we suspect, but it will not by itself tell us whether these planets are Earth-*like*. In other words, it won't tell us if they have atmospheres, oceans, plate tectonics, magnetic fields, or any of the other things that make our planet such a truly great home to life. Now, you already know what I think: I suspect that Earth-size is going to mean Earth-like, at least if the planet has an orbit within its star's habitable zone.

However, while my suspicion probably represents the majority view among scientists, science is not a democracy. For the moment, I have no hard evidence to support my suspicion, which means it's possible that I'm dead wrong. In fact, there are a few scientists who would say that I am, and who have cited a number of reasons why they believe that Earth-like planets will prove to be quite rare. Their idea has been popularized in recent years under the name of the "rare Earth hypothesis."

[4] A European mission called COROT was launched in 2006 and is similarly searching for transits. Although it was expected only to be able to find planets considerably larger than Earth, its early performance is exceeding expectations.

The rare Earth proponents are good scientists, and they have come up with a number of interesting arguments to support their position. However, for every argument they make in favor of their hypothesis, other scientists have come up with counterarguments that go against the hypothesis. Let's look at just a few of the issues that come into play.

To begin with, the rare Earth proponents do not agree with all of the assumptions that led to my earlier conclusion that most stars are capable of having habitable planets in orbit. If you look back at how I reached my conclusion, you'll see that it was based on the fact that all stars have a habitable zone of at least some size around them, and that all but the most massive stars live long enough to allow the possibility of habitable planets forming within that zone. The location and extent of a star's habitable zone depends only on the star's luminosity, so no one argues about this part of it. However, the rare Earth proponents cite additional factors that they think might prevent habitable planets or life from arising around the vast majority of stars, regardless of the sizes we might calculate for their habitable zones. In essence, they argue that there is a *galactic habitable zone*—a relatively narrow ring around the center of our galaxy that is analogous to the habitable zone around an individual star—in which it is possible for a star to have habitable planets, and that no such planets are possible outside this galactic ring.

The arguments for a narrow galactic habitable zone go basically like this: Recall that while all stars are made at least 98 percent or so of hydrogen and helium, they still differ in their proportions of the heavier elements. Because the heavier elements were manufactured by past generations of stars, the heavier element content is lower for older stars (since there had been fewer past generations by the time they formed) and also lower for stars farther from the center of the galaxy (where the greater distances between stars mean less manufacturing has occurred). For the oldest and most distant stars, the fraction of elements heavier than helium can be 100 times smaller than the 2 percent we find in the Sun. This could be important to prospects for life, because terrestrial planets are made almost entirely of these heavier elements (such as iron, nickel, silicon, carbon, and oxygen); indeed, the rare Earth proponents claim that you can get terrestrial planets only if your star is located about as close to the galactic center as our Sun. This part of their argument excludes the possibility of finding Earth-like planets much beyond the Sun's distance from the center of the galaxy. They then rule out regions much closer than the Sun by looking at supernova rates. A supernova is the titanic explosion that ends the life of a massive star, and supernovae are much more common in the more crowded, inner regions of the galactic disk. Because supernovae release tremendous amounts of dangerous radiation and cosmic rays, the rare Earth proponents suggest that Earth-size planets in the inner regions of the galaxy would have their surfaces fried so often by this radiation that life could not get started or survive.

These arguments for a galactic habitable zone make some sense, but there may be ways around them. Having a heavy-element abundance that is 100 times smaller than that of the Sun might indeed seem like a problem, but if you do the math, you'll find that Earth's mass is less than 1/100,000 of the mass of the Sun. Thus, even a very small heavy element abundance might be enough to make one or more Earth-like planets, as long as the lower abundance doesn't inhibit the planet formation process. Regarding the radiation danger from supernovae, we do not really know whether such radiation would be detrimental to life. A planet's atmosphere might protect life against the effects of the radiation. It is even possible that the radiation could be beneficial to life by increasing the rate of mutations and thereby accelerating the pace of evolution. If these counterarguments are correct, then Earth-like planets might be found throughout much or all of the galaxy, with no constraints.

Other rare Earth arguments cite a number of specific features of our solar system and planet that have contributed to the habitability of Earth, and argue that in combination they would be so rare as to be almost impossible to find elsewhere. For example, the rare Earth proponents point out that Jupiter has played a major role in controlling the impact rate in our solar system; without Jupiter, the rate of impacts would have almost certainly remained high throughout geological history, instead of tailing off at the end of the heavy bombardment.[5] They therefore suggest that we are "lucky" to have Jupiter, because otherwise the higher impact rate might have prevented us from ever evolving. However, it's not clear that this is a particularly rare form of luck; our discoveries of extrasolar planets already show that Jupiter-size planets are quite common, and we don't yet know enough to be able to say whether they are also common in Jupiter-like orbits. Moreover, even if the rare Earthers are right in arguing that impacts would be more common in most other planetary systems,

[5] The somewhat subtle reason for this fact is that Jupiter's gravity stabilizes orbits of asteroids in the asteroid belt and, even more important, probably was responsible for pitching billions of icy comets from its region of the solar system to a much more distant region known as the Oort cloud; there, because they are so far away, these comets pose much less of a threat than they would have if they had remained closer to the Sun.

this would not necessarily be a bad thing for evolution: The K-T impact paved the way for our own evolution, so perhaps a planet with more frequent impacts would tend to give rise to intelligence sooner rather than later.

Another set of circumstances that the rare Earth proponents attribute to rare planetary luck are those that keep Earth's climate stable, including plate tectonics (which is part of the climate-moderating carbon dioxide cycle) and the existence of our large Moon (which helps stabilize Earth's axis tilt). Again, however, we can also argue the other way. As we discussed in chapter 6, it's quite possible, though unproven, that plate tectonics are an inevitable feature of any approximately Earth-size planet within its star's habitable zone. Similarly, our Moon's presumed formation in a random giant impact does not automatically mean that large moons should be rare. At least a few giant impacts should be expected in any planetary system. Indeed, Earth may not be the only object in our own solar system that ended up with a large moon because of a giant impact; Pluto's largest moon (Charon) may have formed in the same way (along with its two smaller moons). Moreover, even if Earth's axis swung as wildly as that of Mars, the changes in tilt would probably occur over periods of at least tens of thousands of years—long enough that life might be able to migrate or adapt as the climate changed.

So what should we make of all these points and counterpoints? Easy: At least when it comes to the question of whether Earth-like planets are rare or common, we are no better off than the atomists and Aristotelians more than 2,000 years ago. As I noted when we discussed their ancient argument back in chapter 1, it's a lot easier to argue endlessly when there are no actual facts to get in the way. That is our situation today with regard to the rare Earth hypothesis: We have no way to decide whether it is right or wrong.

But it won't stay this way, and that is the beauty of science. We know exactly what we must do to resolve this debate. We must make the scientific observations that can bring us all to agreement. We must find a way to build telescopes powerful enough not just to detect the existence of Earth-*size* planets, but to actually see and study them in enough detail so that we can tell whether they are also Earth-*like*, and even to tell whether they have life. Remarkably, scientists already have designs on the drawing board for telescopes that could accomplish this goal. Because a few new technologies will need to be developed, and because of the ongoing budget problems faced by science, these telescopes probably won't be built for at least 10 to 20 years. But I'll be very surprised if they don't exist by 30 years from now. At that point, the rare Earth hypothesis will be history, because we'll have the data in hand to know for sure whether it was right or wrong. Personally, I expect that some time around the year 2038, I'll be able to turn on whatever communication device is popular at the time, and hear the news of the discovery of the first Earth-*like* planet around another star. Then, over the next few years, many more such discoveries will pour in, proving once and for all that planets like ours are the rule and not the exception.

BEYOND UFOS

If my guess is correct, and the rare Earth arguments prove to be off-base, then we are back to the staggering numbers I gave you earlier. That is, I would not be at all surprised to learn that there are *100 billion* habitable planets in our galaxy. If just one in 1 million of these is enough like Earth to eventually give rise to a civilization, then we have the 100,000 civilizations that I suggested back in chapter 3. But is it really reasonable to think that intelligence could be this common? And if it is, shouldn't we be able to make contact with some of these other civilizations? These are the questions we'll turn to in the next chapter, as we discuss the search for extraterrestrial intelligence.

Before we go there, however, I'd like to return to the theme with which I opened this chapter. We may or may not share the universe with other intelligent beings, but I believe we can grow up as a civilization either way. I chose the opening quotation from Christiaan Huygens because it so beautifully captures how a new perspective on the universe should help us better understand ourselves. I originally found this quotation in Carl Sagan's *Cosmos,* where it opens his chapter 13. Living three centuries after Huygens, Sagan had the benefit of the far deeper understanding of nature that we have since obtained. Sagan also played a major role in arranging for the *Voyager 1* spacecraft to take its famous portrait of the planets as they appear from the outskirts of our solar system, where our own Earth shows up as just a "pale blue dot" surrounded by what looks a beam of scattered sunlight (figure 8.3). This photograph inspired Sagan to elaborate on the idea expressed by Huygens, and he did this so beautifully that I am compelled to share it with you. So here, excerpted from a commencement address that he delivered just months before dying of cancer in 1996,

Figure 8.3. Earth, photographed from the outskirts of our solar system by the *Voyager* spacecraft. The "sunbeam" surrounding Earth is an artifact of light scattering in the camera. (Courtesy of NASA)

I offer you Carl Sagan's words reflecting on how we might all learn to grow up:

> *We succeeded in taking that picture [from deep space], and, if you look at it, you see a dot. That's here. That's home. That's us. On it, everyone you ever heard of, every human being who ever lived, lived out their lives. The aggregate of all our joys and sufferings, thousands of confident religions, ideologies and economic doctrines, every hunter and forager, every hero and coward, every creator and destroyer of civilizations, every king and peasant, every young couple in love, every hopeful child, every mother and father, every inventor and explorer, every teacher of morals, every corrupt politician, every superstar, every supreme leader, every saint and sinner in the history of our species, lived there on a mote of dust, suspended in a sunbeam.*
>
> *The earth is a very small stage in a vast cosmic arena. Think of the rivers of blood spilled by all those generals and emperors so that in glory and in triumph they could become the momentary masters of a fraction of a dot. Think of the endless cruelties visited by the inhabitants of one corner of the dot on scarcely distinguishable inhabitants of*

some other corner of the dot. How frequent their misunderstandings, how eager they are to kill one another, how fervent their hatreds. Our posturings, our imagined self-importance, the delusion that we have some privileged position in the universe, are challenged by this point of pale light.

Our planet is a lonely speck in the great enveloping cosmic dark. In our obscurity—in all this vastness—there is no hint that help will come from elsewhere to save us from ourselves. It is up to us. It's been said that astronomy is a humbling, and I might add, a character-building experience. To my mind, there is perhaps no better demonstration of the folly of human conceits than this distant image of our tiny world. To me, it underscores our responsibility to deal more kindly and compassionately with one another and to preserve and cherish that pale blue dot, the only home we've ever known.[1]

[1] This quotation from Carl Sagan is an excerpt from a commencement address he delivered on May 11, 1996.

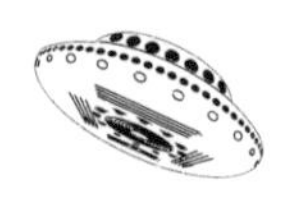

9

THE SEARCH FOR EXTRATERRESTRIAL INTELLIGENCE

Now when we think that each of these stars is probably the centre of a solar system grander than our own, we cannot seriously take ourselves to be the only minds in it all.

—*Percival Lowell (1855–1916)*

Perhaps it seems strange that I would start a chapter with a quotation from Percival Lowell, whose greatest claim to fame comes from having imagined a system of canals and a civilization on Mars that existed nowhere but within his own mind. But his story in many ways parallels the ongoing story that we now find ourselves in. Lowell saw a few real things that seemed to hint at the idea that life might be possible on Mars, such as its polar caps, its seasonal changes in coloration, and the vague surface markings that he mistook for a network of canals. He took these hints so much to heart that he lost his objectivity, and became convinced that he saw not just hints but actual proof. Lowell was a scientist, and in most respects quite a good one, but when it came to life on Mars, he abandoned science and built a case based on faith. Unfortunately for Lowell, his faith was unlike religious faith, for which science can generally say nothing about its validity. Instead, he had a faith that was set from the beginning on a collision course with science, because it was only a matter of time until improved observations would shatter the mythology he had created.

Today, we understand a great deal about the nature of our universe and the stars and planets within it, and this understanding gives us good reason to think that life—including intelligent life—might be possible on many worlds. The idea seems so eminently reasonable that it's tempting to make the same leap of faith as did Lowell, and conclude not just that other civilizations are *possible*, but that they really exist. Indeed, a great many people have already has made this leap, and fervently believe in the UFOs and alien visitation that I discussed earlier in this book. Scientists, perhaps taking the lesson of Lowell to heart, have been far more cautious: I know many astrobiologists who will say that their best guess is that other civilizations exist, but I've never heard one claim to know it for certain. However, a few scientists have somewhat oddly taken the opposite leap of faith, using a variety of arguments (most of which go far beyond the well-reasoned arguments of the rare Earthers) to conclude that we must be alone in the universe.

Some of you reading this book might fall into the faith-based camps. Perhaps some of you have seen your own UFO, and hence are convinced that other civilizations exist. Perhaps others of you have read elsewhere about the so-called "anthropic" arguments and have become convinced that no one else is out there. In either case, I wish you luck, because like Lowell's beliefs a century ago, your beliefs are on a collision course with science. One way or the other, we will eventually find out whether other civilizations are common. It may take decades, or even centuries, but unless we stop searching, we will find an answer.

Now, you may be thinking that I have too much faith in science, and that an answer could perhaps elude us indefinitely. But I really don't think it can. Remember, we are searching for others *like us*. We are not looking for those who might blend into the forest like frogs or even monkeys, or live hidden in the depths of the ocean, or lie unseen in rocks beneath the ground. We are looking for advanced beings who have built a civilization, and the building process must always leave indelible marks on any world. Someday, we will have telescopes powerful enough to *see* these marks, if they exist. We may see their city lights on the night side of their world, or discover the spectral signatures of chemicals such as CFCs that do not exist naturally and could only be the products of a civilization.

It will probably be a while before our telescopes are this powerful, but we may not even have to wait that long. If other civilizations are truly like us, they will have science, and they will discover the laws of nature, just as we have. They will learn to build radios and televisions, so that they can communicate with each other at long distance around their planet. And at that point, like us, they will be broadcasting their existence to anyone who wants to learn of it.

It's true: We cannot hide anymore. For more than seven decades now, we have been broadcasting radio and television transmissions that go beyond Earth's atmosphere and out into space. Once in space, they travel outward at the speed of light. With sufficiently powerful radio telescopes, anyone within about 70 light-years of Earth could now be watching our old TV shows—a somewhat scary thought that might explain why no one wants to visit us. A hundred years from now, our signals will have reached out to 170 light-years from Earth. A hundred thousand years from now, everyone in the galaxy will, at least in principle, have had a chance to learn that we were here, even if we are by then long gone. In that sense, we have already gone beyond making an indelible mark on our planet, for we have made one on the universe.

The scientific search for extraterrestrial intelligence—SETI—is in essence a search for such marks on the universe. SETI researchers use radio telescopes, and sometimes other types of telescopes as well, to search for signals that others might have broadcast into space. There's no guarantee of success, because we do not yet know if anyone is out there. But as two of the pioneers of SETI (Philip Morrison and Giuseppe Cocconi) wrote in 1959, "The probability of success is difficult to estimate; but if we never search, the chance of success is zero." It is time for us to examine the issues behind the search for extraterrestrial intelligence.

THE DRAKE EQUATION

We do not know whether or how many other civilizations exist, but it can be helpful to define our ignorance. That is, instead of just saying "we don't know," we can try to figure out exactly what it is that we don't know. While this type of exercise still doesn't answer the question, it can give us guidance about how we might go about finding the answer in the future.

The first and most famous effort to define our ignorance about extraterrestrial intelligence was made by astronomer Frank Drake. In setting the agenda for a 1961 scientific meeting about the search for extraterrestrial intelligence—the first meeting of its kind—Drake decided to try to summarize the factors that would determine whether attempts to detect intelligent extraterrestrials could succeed. In doing so, he wrote down a simple equation, now famous as the *Drake equation,* that at least in principle could be used to calculate the number of civilizations existing elsewhere in our galaxy or in the universe at large.

Before I show you the equation, two important notes are in order. First, don't panic: the equation involves nothing more than multiplication, so even those of you who are math-phobic can understand it. Second, we need to define "civilization" in this context. Because Drake was considering the possibility of receiving a radio signal from other beings, his equation was designed to calculate the number of civilizations *that are capable of broadcasting signals into interstellar space*. This definition clearly suits the purpose of SETI, though it is somewhat different from the definition of civilization that we use in everyday life. For example, the ancient Greeks don't count under this definition, because they never developed radio or other technologies that could be used to send messages to the stars.

Drake's original equation separated out seven distinct factors that would contribute to the number of civilizations; I imagine that quite a few of you readers will have seen it written elsewhere in its original form. However, some years ago as I was teaching the equation to students in my introductory astronomy classes, I found that I could help them understand it better by combining some of Drake's individual terms. In addition, new discoveries since 1961 have made a few of Drake's original assumptions somewhat out of date. For example, he started with a term based on the *rate* of star formation in a galaxy, which made sense in 1961, when it was hypothesized that star formation would proceed at a steady rate at all times. We have since learned otherwise; for example, the galaxies known as *elliptical galaxies* have little or no present-day star formation, so the rate term in Drake's original equation would imply that they could not have any civilizations at all. Today, we assume that if there are stars there may be planets, and if there are planets there may be habitable planets and life. As a result of these changes, the version of Drake's equation that I presented in class had only four terms instead of seven, and they are a bit different from his original terms. Now, I hope you know that I try to be true to the science both in writing and teaching, so I did not take messing with a famous equation lightly. You can therefore imagine my relief when Drake himself gave his blessing to the altered version.[1] So here, in its slightly modified form, is the Drake equation:

$$\text{Number of civilizations} = N_{\text{HP}} \times f_{\text{life}} \times f_{\text{civ}} \times f_{\text{now}}$$

See? I told you it would just be multiplication. Let's go through the terms to make sense of it. N_{HP} is the number of habitable planets in a galaxy; here, we are referring to planets with *surface* habitability—meaning the presence of liquid water and adequate sunlight—so this term represents the number of planets that are at least capable of having Earth-like

[1] I have not had the opportunity to meet Drake myself. However, the co-author of my *Life in the Universe* textbook, Seth Shostak, knows him quite well, and both are affiliated with the SETI Institute. Seth presented Drake with the rationale for our using the modified version for our textbook, and Drake agreed.

life. The next term (f_{life}) is the fraction of habitable planets that actually *have* life of some kind, meaning microbial life. For example, if $f_{life} = 1$ it would mean that all habitable planets have life; if $f_{life} = 1/1{,}000{,}000$ it would mean that only 1 in a million habitable planets has life. The product $N_{HP} \times f_{life}$ therefore tells us the number of life-bearing planets in the galaxy.

The third term (f_{civ}) is the fraction of the life-bearing planets on which evolution has proceeded to the point where a civilization capable of interstellar communication *has at some time* arisen. For example, if $f_{civ} = 1/1{,}000$ it would mean that such a civilization has existed on 1 out of 1,000 planets with life, while the other 999 out of 1,000 have not had a species intelligent enough to build radio transmitters, high-powered lasers, or other devices for interstellar conversation. When we multiply this term by the first two terms (to form the product $N_{HP} \times f_{life} \times f_{civ}$), we get the total number of planets on which intelligent beings have evolved and developed a civilization at some time in the galaxy's history.

Finally, f_{now} is the fraction of the civilization-bearing planets that happen to have a civilization *now*, as opposed to, say, millions or billions of years in the past. This term is important because it tells us how many civilizations we could potentially get a signal from, because civilizations that are long gone are no longer broadcasting.[2] Because the previous three terms told us the total number of civilizations that have *ever* arisen in the galaxy, multiplying by f_{now} tells us how many civilizations we could potentially make contact with today. For example, if the first three terms were to tell us that 10 million planets in the galaxy have at some time had a communicating civilization but f_{now} turns out to be 1 in 5 million, then only two civilizations would be expected to exist today. As we will see shortly, the value of f_{now} depends on how long civilizations survive once they arise.

The Drake equation helps to define our ignorance because, even though we don't know the values of any of its terms, we have more information about some terms than others. To keep things simple, we'll look at the values only within our own Milky Way Galaxy; it would be straightforward to extend the results to the rest of the universe.

The first term (N_{HP}) is probably the most constrained. Based on our discussions in chapter 8, we expect the number of habitable planets to be quite

[2] For the purposes of the Drake equation, we assume that the term f_{now} takes into account the light-travel time for signals from other stars; for example, if a star with a civilization is 10,000 light-years away, it counts in determining f_{now} if the civilization existed 10,000 years ago, because signals broadcast at that time would just now be arriving at Earth.

large—perhaps as large as 100 billion or more—*unless* the rare Earth arguments are correct. Since I'm personally skeptical of these arguments, I'm going to take 100 billion as a fair estimate of N_{HP}. The second term (f_{life}) is far less well understood. However, if life really does get started as easily as it seems it did on Earth (see chapter 5), then we'd expect this term to have a value close to 1, meaning that almost all habitable planets would actually have life on them. It therefore seems reasonable to imagine that our galaxy has 100 billion life-bearing planets, leaving the last two terms to whittle down the number of planets that have a communicating civilization. These two terms are even less well understood than f_{life} and are therefore worth a bit more investigation.

THE QUESTION OF INTELLIGENCE

Even if life-bearing planets are very common, civilizations capable of interstellar communication might not be. The fraction of life-bearing planets that at some time have had such civilizations (f_{civ}) depends on at least two things: First, a planet would have to have a species evolve with sufficient intelligence to develop interstellar communication. In other words, the planet needs a species at least as smart as we are. Second, that species would have to realize its potential by actually developing a civilization with technology as advanced as ours.

Although we cannot really be sure, most scientists suspect that only the first requirement is difficult to meet. A fundamental assumption in nearly all science today is that we are not "special" in any particular way. We live on a small planet orbiting an ordinary star in a normal galaxy, and we assume that living creatures elsewhere—whether they prove to be rare or common—would be subjected to evolutionary pressures quite similar to those that have operated on Earth. Thus, if species with intelligence similar to ours have arisen elsewhere, we assume that they would have similar sociological drives that would eventually lead them to develop the technology necessary for interstellar communication.

If this assumption is correct, then the fraction f_{civ} depends primarily on the likelihood of sufficient intelligence evolving on a planet that already has microbial life. As we discussed in chapter 5, the fact that it took some 3 billion years for life on Earth to go from the microbial to even the most primitive animals suggests that this step might be quite difficult. Nevertheless, if we continue with our assumption that Earth is "typical," then it would seem reasonable to find animal life on many worlds that are as old as

Earth. Because most of the stars in the galaxy are actually older than the Sun, this constraint would barely affect our overall estimates.

Just getting animals is still not enough, however; we need really smart animals, like us. As I mentioned briefly in chapter 5, my personal guess is that once you get animals, you'll eventually get smart ones. That's not to say that there aren't evolutionary dead-ends: The dinosaurs, for example, spent some 180 million years evolving and diversifying as the animal kings of the planet, yet as far as we know, none of them ever got clever enough to discover fire or build a wheel, let alone to build radio telescopes or spacecraft. Perhaps they might have if they'd just had more time, but that's another reason why I suspect that intelligence may be inevitable: Because dinosaurs didn't develop intelligence quickly enough on their own, nature found a way of getting them out of the way—with the K-T impact—and allowing the mammals to give it their own college try. But I'm just guessing; is there any *evidence* that we can use to look at the likelihood of animals becoming highly intelligent?

Yes, although like most things in this field of astrobiology, the data can lead to conflicting interpretations. The necessary data are measurements of animal intelligence, for both past and present animal species. Intelligence is somewhat subjective, but we generally presume that it is related to brain size (across species, not between individuals of a species). However, it can't be brain size alone, because larger animals tend to have larger brains regardless of their "smarts"—you need more brain power to control a bigger body, regardless of whether you also use that brain for intellectual pursuits. Scientists who study animal intelligence have therefore come up with a measure called the *encephalization quotient*, or EQ, which is based on the ratio of brain mass to body mass. The larger the brain mass in proportion to the body mass, the smarter the species.

Although the data are sparse, studies of extinct species suggest that there has indeed been a general upward trend in animal intelligence, at least as measured by EQ, over the past couple hundred million years. This makes sense, because intelligence is clearly a beneficial evolutionary adaptation, especially among predators: It makes them better able to catch prey and to avoid other predators. But while we'd expect natural selection to select for intelligence that increases a predator's survival prospects, it's not clear why natural selection would lead to brains big enough to build radio telescopes and spacecraft. Some anthropologists have advanced hypotheses that might explain why this would be an evolutionary tendency, but others suspect we just got lucky to end up with such big brains in proportion to our body sizes. Here, the EQ data offer an interesting perspective.

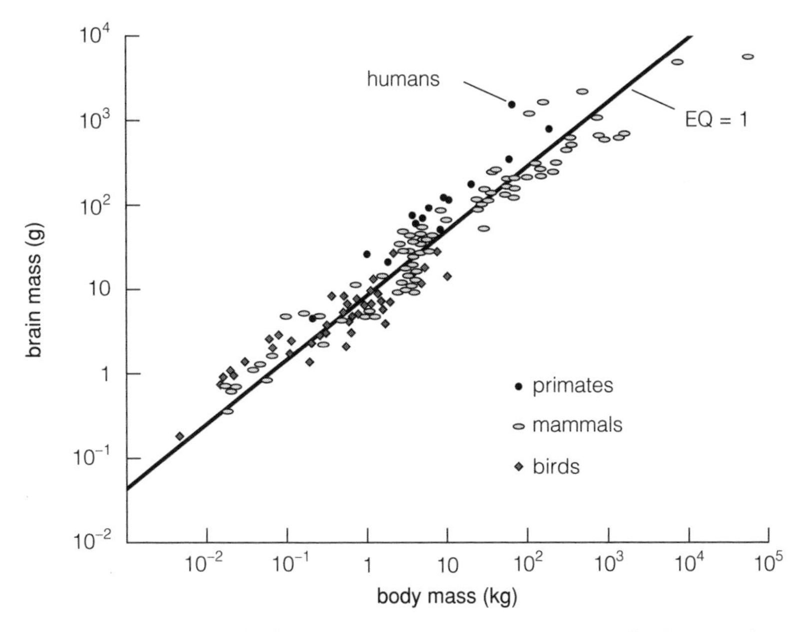

Figure 9.1. This graph shows how brain mass compares to body mass for various mammals (including primates) and birds. The straight line represents an average of the ratio of brain mass to body mass, so that animals that fall above the line are smarter than average and animals that fall below the line are less smart. Note that the scale uses powers of 10 on both axes. (Adapted from a 1977 article by Carl Sagan.)

Figure 9.1 shows EQ data for a sampling of animal species living today. The straight line represents the "average" EQ: Animals whose brain mass falls above the line are smarter than average, while animals whose brain mass falls below the line are not. Keep in mind that it is the *vertical* distance above the line that tells us how much smarter a species is than the average, and notice that the scale goes in powers of 10 on both axes. If you look closely, you'll see that the data point for humans lies significantly farther above the line than the data point for any other species. The conclusion: As measured by EQ, at least, we are *by far* the smartest species that has ever existed on Earth. In fact, by this measure, we are nearly twice as smart as dolphins, the animals with the next highest EQs.

To some extent, these data support the idea that we got "lucky" to end up with such big brains. After all, there are millions of other animal species, and we ended up being not only smarter but a lot smarter. However, the same data can also be used to reach an opposite conclusion. The scatter in the levels of intelligence among different animals tells us that some variation should be expected, and statistical analysis shows that we are not unreasonably far above the average. We are the first species to have a large enough deviation above the mean to build spacecraft, but if we destroy ourselves, we might not be the last. From this point of view, my guess that civilization will eventually follow on any planet that gets animals seems at least somewhat grounded in reality.

TECHNOLOGICAL LIFETIMES

For the sake of argument, let's assume that life and intelligence are indeed at least reasonably likely, so that thousands or millions of planets in our galaxy have at some time given birth to a civilization. In that case, the final factor in the Drake equation (f_{now}) determines the likelihood of there being someone out there with whom we could make contact. The value of this factor depends on the survivability of civilizations.

Consider our own example, and to keep the numbers easy let's suppose that it has been possible, at least in principle, for civilizations to arise any time in the past 7 billion years (a reasonable assumption, since the universe was by then already 7 billion years old). As noted earlier, we have been capable of interstellar broadcasts via radio for about 70 years. So if we were to destroy ourselves tomorrow, our technological "lifetime"—the length of time during which we were able to make ourselves known to other star systems—would have been only 70 years, or 1 part in 100 million of the time during which civilizations have been possible. If this technological lifetime were typical of other civilizations, and if we assume that the times at which other civilizations have emerged is randomly distributed over the past 7 billion years, then we would expect only 1 in 100 million of the civilizations that have ever existed to be out there right now. In that case, there would have to have been at least 100 million civilizations during our galaxy's history for there to be a decent chance that even one other civilization exists now. Our SETI efforts would come to naught, because we'd be listening for the sounds of silence.

Of course, we have not yet destroyed ourselves, so it's possible that our technological lifetime will be a lot longer, which would in turn give us reason to think that the fraction f_{now} could be much higher. For example, and again to keep the numbers easy, suppose civilizations routinely survive for

700 million years. Then given 7 billion years during which civilizations could have arisen, the random chance of a civilization surviving now would be one in ten. In that case, even if only 100 civilizations had ever arisen, we'd expect there to be ten of them out there right now, potentially accessible to our SETI searches.

The critical conclusion, first recognized by Frank Drake himself, is that the survivability of civilizations is probably the key factor in whether any are out there now. If most civilizations self-destruct shortly after achieving the technology for interstellar communication, then we are almost certainly alone in the galaxy at present. But if most survive and thrive for thousands or millions or billions of years, the Milky Way may be brimming with civilizations—with most or all of them far more advanced than our own.

CONDUCTING THE SEARCH

If there are indeed other civilizations using radio or other forms of light for communication, then it seems at least worth listening for them. That is precisely what most SETI researchers do: They use large radio telescopes—and occasionally other types of telescopes in case the aliens might be signaling with, say, lasers rather than radios—in hopes of receiving a broadcast signal.

To some degree, SETI efforts have been ongoing since 1960, when Frank Drake undertook the first formal effort to search for signals that might be coming from civilizations orbiting two nearby stars (Epsilon Eridani and Tau Ceti). Searches have been sporadic, however, both because of the limited amount of observing time available at the world's large radio telescopes, and because, at least for the past fifteen years, our government has not seen the wisdom of spending money on SETI. Fortunately, private donors have stepped up to the plate, and SETI researchers are now in the process of building the world's first array of telescopes dedicated to searching for signals from other civilizations. This *Allen Telescope Array*—named for Paul Allen, who donated much of the money for the project—has already begun limited operations (color plate 8a). When completed, the Allen Telescope Array will consist of 350 radio dishes, all working together to listen for alien communications.

The Allen Telescope Array will greatly improve our ability to search for signals, but it's important to realize that our technology remains limited. In particular, we are nowhere near the point of being able to pick up the kinds of radio and television signals that we ourselves broadcast, even if they are coming from the nearest stars. The problem is sensitivity: Because radio and television signals are not beamed in any particular direction, they spread out in all directions in space. As a result, they become much weaker with distance, and we probably won't be capable of detecting such signals even from a few light-years away until we can build extremely large radio telescopes in space or on the Moon.

So what are the SETI scientists searching for? The answer is that, with current technology, they are searching for strong signals that someone might have *deliberately* beamed in our direction. In other words, for the time being at least, we search with the hope that someone wants to make their presence known. This certainly limits the possibilities of detection much more than if we could search for "accidental" interstellar broadcasts like those of radio and television, but it's plausible to imagine that deliberate signals are out there.

Again, in assuming that we are not in any way special, we are also led to assume that other civilizations would share our interest in making contact. In that case, they might decide to build large transmitters that could beam powerful signals in the direction of stars that seem to offer the promise of hosting life. For example, they might choose to send transmissions toward our solar system after having discovered the presence of planets within it, and perhaps even having recognized that our system has a planet of habitable size orbiting within the habitable zone. Alternatively, they might have a set of transmitters, each capable of sending out a powerful beam that would cover millions of star systems; in that case, they wouldn't need to know anything in particular about our solar system and we might still receive their signal.

We ourselves have already sent a few such transmissions: For example, in 1974, scientists used a powerful radar transmitter on the Arecibo radio telescope (located in Puerto Rico) to aim a three-minute broadcast at the globular cluster M13. This cluster contains a few hundred thousand stars, seemingly offering a decent chance that at least someone within it might happen to have a radio telescope pointed in the right direction when the three-minute transmission passes by on its outward journey into the universe. Note, however, that M13 is located about 21,000 light-years away, so it will take some 21,000 years for our signal to get there and another 21,000 years for any response to make its way back to Earth. It's not a convenient method of two-way conversation, but if someone there ever receives our signal, it's possible that it will represent their first absolute proof that they are not alone in the universe.

This idea leads to one more question that many people have about the search for extraterrestrial intelligence: How would they know—and how would we know—that a signal really comes from another civilization and is not just a naturally occurring bit of radio static from the stars? For "accidental" broadcasts such as radio and television shows, it might be difficult to decode them without detailed study. Remember, however, that our current SETI efforts could detect only deliberate signals. If someone is going to the trouble of deliberately trying to make their presence known, they'd presumably also make it as easy as possible for someone else to distinguish their signal from natural static. For example, our 1974 signal consisted of precisely 1,679 "bits" (in this case, tiny frequency changes)—which any civilization that has discovered mathematics will recognize as the product of the prime numbers 23 and 73. Stars and other nonliving things don't know about prime numbers, so this and other strategies should be able to make an intelligent civilization "obvious" to other, similarly intelligent civilizations.

BEYOND UFOS

Some people are worried about the prospects of radio signals beaming among the stars—worried not about our hearing from someone else, but about someone else hearing from us. They worry that, in beaming out messages like the one from 1974, we are exposing ourselves to danger. If Earth-like planets are rare, they reason, letting someone know that we have one here could be an invitation to come try and steal it. Better to stay quiet, they suggest, at least until we can be sure that aliens would not mean us any harm.

If this concern has ever crossed your mind, I have some advice for you: Relax. As we discussed in chapter 3, anyone capable of traveling from their home star to here would be so far beyond us technologically that there's no chance at all that some deliberate broadcast—whether the one in 1974 or new ones that we might beam out in the future—will be the first that they hear from us. If they're on their way with hostile intentions, it's because they've seen our television shows, not because we sent out a friendly "hello." And lest that thought cause you even further worry—doesn't everyone want to "kill your TV" at some point?—remember that we pose no threat to them, and we have nothing that they'd need. If they can travel this far, they are at least thousands of years, and possibly millions or billions of years, ahead of us in technology and science. We can't hurt them, and there's nothing Earth has to offer that they can't manufacture much closer to home.

In fact, I rather doubt that any such advanced aliens would be paying any attention to us at all, except perhaps for monitoring us, waiting to see if we ever prove ourselves smart enough and friendly enough to deserve an invitation into their galactic club. But that's a story for the next chapter.

No matter how advanced a civilization may be today, they must have gone through a less advanced stage, when they would have been like us, listening for others and broadcasting their own existence. And no matter how advanced they may have become since then, they might still be signaling in a way that is detectable to us, perhaps because the laws of physics do not allow them to hide, or alternatively because they *want* us to be able to learn they are out there, once our technology is sufficiently powerful.

In that sense, SETI is a bit like a wing and a prayer. We do what we can to catch a signal from someone else, and hope that the signals are indeed out there. With the new Allen Telescope Array, our chances are better than ever, but still impossible to predict. Of course, that doesn't stop people from trying. My friend and colleague Seth Shostak, who has been deeply engaged in SETI research for decades and is also the co-author of my *Life in the Universe* textbook, believes we will get a signal within the next 20 to 30 years. I'm somewhat less optimistic, but I've got to admit, I sure hope he's right.

10

WHERE IS EVERYBODY?

Sometimes I think the universe is full of life and sometimes I don't. Either way is equally amazing.
—*Arthur C. Clarke*

I began this book by telling you that we live in a universe that contains worlds beyond imagination, and I have now spent nine chapters explaining why it seems likely that many of those worlds should be inhabited, some with beings like us. I have also explained why—although I consider it at least remotely possible that some UFOs could indeed be spacecraft from distant worlds—most claims of alien visitation do not make sense once you realize how advanced such aliens would have to be. But if civilizations really are as common as it seems they ought to be, shouldn't we by now have some *real* evidence of their existence? Or, as it was simply put in 1950 by the Nobel-prize winning physicist Enrico Fermi, during a conversation with other scientists who were speculating about extraterrestrial intelligence, "So where is everybody?"

This seemingly innocent question, which has come to be called the *Fermi paradox*, turns out to be surprisingly difficult to answer. Moreover, the deeper we look into the issues that underlie the question, the more profound its implications become. This question will therefore be the topic of this final chapter, but let me tell you now where we'll end up: I intend to show you that, if and when we learn the answer to the "Where is everybody?" question, it will cause the most dramatic shift in the status of our human species that has ever occurred in history.

I realize that this must sound outlandish, which is why I've told you where we're headed in advance. After all, if I want you to believe me, not only do I have to make my case, but I must do so in such a way that you don't reach the end and say "no way." So that's why I've prepared you up front, and also why I name-dropped a Nobel prize winner: Fermi was not being glib; he, too, recognized the deep implications of the question that he asked. Many others have done the same and, indeed, I chose to open this chapter with the quotation from Arthur C. Clarke for just that reason: His words make a key point, which is that it's not so much *what* the answer to the question turns out to be that is important, but the mere fact that it must *have* an answer, one way or other.

HOW CIVILIZATIONS CAN GROW UP

A couple of chapters back, I offered you words from Christiaan Huygens and Carl Sagan, each explaining how new perspectives on our place in the universe should help us grow up as a civilization. But we have not grown up yet, a sad fact that we are reminded of everyday in the news, as we read about terrorists, hatred, wars, and abject poverty. A grown-up civilization would have learned to do better.

In fact, there's no guarantee that we'll ever grow up. We constantly discover new ideas and develop new technologies that could make the world a better place, but we seem as likely to put them to work for destructive as for constructive ends. Sometimes, when I'm feeling down, I despair that as a species, we just don't care enough to realize our potential, and that centuries from now, archaeologists will sift through the ruins of our civilization and wonder what went wrong. In even deeper moments of angst, I fear that we'll do so much damage to our planet that we'll go the way of the dinosaurs, and it will be millions of years before the Earth sees another set of intelligent beings. In these moments, I think of the art, the music, the dance, the literature, the sports, the science, and the other great things that humans have created . . . and I'm overcome with sadness at the thought that all would be lost forever.

I share these unhappy thoughts because I think they are important for everyone to contemplate. We need some global guilt. We need for everyone to look at the faces of children, and think about how we'll feel if they grow up in a world in which our civilization is collapsing because we, as individuals and as a society, made the wrong choices. Sometimes, I picture future generations looking back at us, putting us on trial, and judging us for our sins. But then I remember that if we don't change, if we don't learn to grow up, there may be no future generations. There will be no one left to judge us—except perhaps God, who surely would not be pleased—so we must judge ourselves. I think if we all take a hard look at our society today, we'll judge ourselves failures, not because we haven't done a lot of things right, but because we still do too many things wrong. It's only once we recognize our failures that we'll be able to turn them around, and prove ourselves worthy stewards of the incredible good fortune that we have inherited from generations past on this remarkable planet.

As I've said from the beginning, this is a book about possibilities, and the possibility of our own demise is frighteningly real. But it is not the only possibility. It is also possible that we *will* grow up, and that other civilizations have grown up too. So part of understanding the "Where is everybody?" question means thinking about what a grown-up civilization might do.

We could debate this issue endlessly, because it is another of those for which we have no facts upon which to base a conclusion. We can't know for certain what a grown-up civilization would be like, because we have never met one and we don't yet qualify ourselves. Nevertheless, I think I have a pretty good idea of what a grown-up civilization would do: It would keep on growing, not necessarily in numbers, but in wisdom. After all, the individuals we admire most are those who spend their lives learning, questioning, and growing wiser. I think the same will prove true for civilizations.

It is possible to argue (and some people do) that wisdom would mean turning inward, looking for meaning only on the home planet. But I don't think so. Introspection is a part of wisdom, but so is looking outward. Indeed, I believe that a civilization can overcome its worldly problems and grow up only by embracing what lies beyond, not by neglecting it, which is why I believe so strongly that we must continue to explore our universe, both with robotic instruments and with people. I see the return of people to the Moon, and the laying of groundwork for future journeys to Mars and beyond, not as luxuries but as necessities.

So if you don't mind, I'll climb back on my soapbox one last time, and tell you what I think we must do if we ever hope to grow up. We must be honest, admitting our failures and finding ways to redress them. We must use our know-how to find solutions to problems such as global warming, debilitating disease, terrorism, poverty, and war. We must use our compassion to teach all people to respect all others, regardless of their ethnicity, religion, or gender. And we must continue to explore, because exploration allows us to dream of what awaits us if and when we accomplish all these other things.

I believe that we can accomplish all these things only if we do them all together, not one at a time. Sure, it costs money to launch telescopes into space, to send robotic probes to Mars and Europa, and to send people back to the Moon. But I think it is money well spent. For those who say that we need to deal with our problems at home first, I ask you to remember that we are an adolescent civilization. Like an individual teenager, we are struggling to make our way to adulthood, and not always behaving in ways that serve our best interests. Just as you don't solve a troubled teen's behavior problems by locking her in a closet and telling her that she can come out when they're fixed, we won't solve our societal problems if we turn only inward. If you want to change the teen's behavior, then in addition to addressing her immediate problems, you must also find a way to inspire her to the point where she herself believes that the problems are worth fixing and that she has a great future ahead. In precisely the same way, I believe that we can grow up as a society only when every individual person, in every individual nation, grows up with enough inspiration to care about making this world a better one.

In my opinion, there's no better way to provide the necessary inspiration than to build a permanent, large, and international research station on the Moon. A lunar base could offer many direct benefits. The new technologies we'd develop both for building it and for operating it would surely have applications here on Earth, as would the discoveries we could expect to make through space-based research. For astronomers and astrobiologists, the Moon could be used as a platform upon which to build giant telescopes that might answer many of our deepest questions about the universe, and that would allow us to begin our studies of planets around other stars. It's even possible that the Moon would have some resources to offer (such as helium-3) that might by themselves be valuable enough to justify the cost of going there. However, by far the most important reason for building the Moon colony is because it would be a place where people of every nation, every race, and every religion could be working together in pursuits to advance the common good.

In fact, I place a much higher value on building the Moon base than on sending people to Mars, at least for the short term, because the Moon is so much more recognizable in the sky.[1] All of us, no matter where we live, can see the Moon, so a Moon base will be a concrete reminder of what we are

[1] Another pertinent argument concerns space tourism, which I support not only because I'd like to go myself but also because it will lead to commercial ventures that should help bring down the cost of space exploration. Tourist trips to the Moon are within reason: It only takes about two days to get to the Moon, so a two-week vacation could easily accommodate a lunar visit. In contrast, until we have rockets far more advanced than today's, Mars is reachable only when it is nearly aligned with Earth on the same side of the Sun—and because that occurs only every 26 months, trips to Mars will require at least two years away from home. (Radiation exposure on the long Mars trip may also be a problem.)

capable of when we are at our best. Imagine a world in which even the poorest children can look up at the Moon and know that people just like them are living and working there. Imagine a world in which Arabs and Israelis, Chechens and Russians, Americans and Iraqis, can all look up and say, "We are working together up there, so surely we can work together down here." If we want to grow up, this is not just the world we must imagine; it is the world we must build.

What we must do is also what other civilizations must do, as long as we maintain our assumption that we are not "special" in any way. I can't prove it, but I'm personally convinced that any grown-up civilization will have built its moon base (if its planet has a moon), visited nearby worlds in its star system, and, ultimately, learned to travel to the stars. But that brings us back to Fermi's question.

As I explained in chapter 3, interstellar travel would be a tremendous challenge. A civilization might colonize its entire planetary system, and still be far from having the technology to move on to other stars. Nevertheless, if the civilization survives long enough, time is on its side. It may take a thousand years from the time of the first space journeys to develop the technology to reach the stars, but this is a short time in the life of a planet or a star. We can go through the challenges of interstellar travel in any number of different ways, but the final conclusion always comes out the same: If other civilizations exist, and if they have had the will to explore, then by now they should be traveling among the stars.

Perhaps they've done it by learning to build spacecraft that travel at speeds near the speed of light, where the mind-bending effects of Einstein's theory of relativity mean that the travelers could make trips in much less time than the time that passes on their home worlds. Perhaps they've done it by developing technologies that we now envision only in science fiction, such as passages through wormholes or hyperspace. But even if none of these advanced types of spacecraft are possible, a civilization could still visit the stars just by finding a way to accept long journeys. Even with our own current technology, we could in principle build spacecraft that would reach nearby stars within a couple of centuries. Such trips might be feasible if we developed ways to hibernate during the long trips, or if we built spacecraft large enough to allow multiple generations during the journey. Long trips might even be made possible through medical science: If we found a way to extend our lifetimes to thousands of years, trips lasting centuries might be no big deal.

To me, there is no question that if we ourselves survive, and if we do not lose our will to explore, we will someday move outward from our home star system, and begin to colonize the galaxy. And in a galaxy that predates our solar system by nearly 10 billion years—in which, as we discussed in chapter 3, we might expect the next youngest civilization to be some 50,000 years ahead of us, and the third youngest to be 100,000 years ahead of us, and so on—it therefore seems that others should already have colonized the galaxy by now. So where is everybody?

POSSIBLE SOLUTIONS TO THE PARADOX

I think you can now see why the paradox runs so deep. It really does seem like others should be out there, all around us, and yet we don't have any evidence that this is the case.

Since the time that Fermi first posed the question, many people have advanced possible solutions to the paradox. Depending on where you look, you can find lists of potential solutions laid out in different ways and with different emphasis. But when I consider them myself, I find that I can group all of them into just three major categories:

1. *We are alone.* There is no galactic civilization because civilizations are extremely rare—so rare that we are the first to have arisen on the galactic scene, perhaps even the first in the universe.
2. *Civilizations are common, but no one has colonized the galaxy.* If the first solution is incorrect, then we are back to the idea that civilizations are common, which takes us back to the types of numbers that we discussed earlier in the book. In that case, we almost inevitably conclude that there should have been thousands, and probably tens of thousands, of civilizations before us in our galaxy alone. This second solution holds that this has indeed been the case, but that not one of these civilizations has gone on to travel among the stars.
3. *There IS a galactic civilization,* but we have so far been unable to discover its existence.

With these potential solutions to the paradox before us, I can begin to show you why the answer to the "Where is everybody?" question has such profound implications to our species.

Let's start by thinking about the implications of the first solution—that we are alone. If this is true, then our civilization is all the more remarkable. The "we are alone" solution implies that through all of cosmic evolution, among countless star systems, we are the first piece of our galaxy or the universe ever to know that the rest of the universe exists. Through us, the universe has attained self-awareness. Some philosophers and many religions argue that the ultimate purpose of life is to become truly self-aware. If so, and if we are alone, then the destruction of our civilization and the loss of our scientific knowledge would represent an inglorious end to something that took the universe some 14 billion years to achieve. From this point of view, humanity becomes all the more precious, and the collapse of our civilization would be all the more tragic. Knowing this to be the case might therefore help us learn to put petty bickering and wars behind us so that we might preserve the universal self-awareness that we as a species represent.

The second solution has much more terrifying implications. If thousands of civilizations before us have all failed to achieve interstellar travel, what hope do we have? Remember, the idea that we are not "special" is really much more than an assumption, because it is based on the fact that the same laws of nature operate throughout the universe as here on Earth. Unless we're missing something major in our understanding—something that would make our own desire to explore highly unusual—then it seems reasonable to assume that other adolescent civilizations would at some point begin building spacecraft just as we have, and that, like us, they would ultimately seek to explore the stars. In that case, the most likely explanation for the absence of a galactic civilization would be that essentially all other civilizations have destroyed themselves before reaching the point at which their technology could take them to the stars.[2] It must be *all* rather than just some, because given the long time that the galaxy has been around, even a single surviving civilization probably would have had enough time to have colonized the entire galaxy by now. (Even with technology only slightly greater than our own, models show that a civilization could in principle colonize the entire galaxy in no more than a few tens of millions of years—which means in considerably less than 1 percent of the galaxy's current age.) This second solution therefore has ominous implications to us, because it would mean that if we are to survive, we probably must beat odds that no one else has ever beaten. It would not bode well for our future, but if it is true, at least it would tell us how much harder a task we face in securing our own survival.

[2]One reviewer of this manuscript pointed out another possibility: Perhaps they find ways to engage in "virtual travel" to the stars, in which they can explore without physically leaving their planet. Given the tremendous growth rate of computing power, this could well be possible. However, I don't think it requires a separate category of solution, since in doing so they would presumably still become aware of other civilizations and hence could "join up" into the galactic civilization that I call the "third solution" to Fermi's paradox.

Now, before we discuss the third solution, we should take a slightly harder look at the first two. Some people find the "we are alone" solution to be appealing; the rare Earth proponents would even say that it is likely. But remember the numbers that I gave you in chapter 1: Just to *count* 100 billion stars—a low estimate of the actual number in the Milky Way Galaxy—would take you more than 3,000 years. When we extend the numbers to the universe, we find that the total number of stars in our universe is comparable to the total number of grains of sand on all Earth's beaches, put together. Does it really seem reasonable to imagine that among all those grains, ours is the only one that sparkles with intelligent life? I'll grant that it's possible, but it sure seems difficult to believe (color plate 8b). If I were placing bets, I'd bet heavily against the "we are alone" solution to Fermi's paradox.

A harder look at the second solution makes us wish it would go away. After all, based on our current example, the idea that adolescent civilizations *always* destroy themselves is depressingly easy to believe. While it's always possible that we could become the first civilization to beat the long odds, I'd feel a lot better about our chances if I knew that others had already made it. So all I can really say about the second solution is that while it seems possible, and perhaps even likely, I *hope* it's not correct.

Perhaps you are starting to understand what I told you at the beginning of this chapter: that the answer to the "Where is everybody?" question will cause the most dramatic shift in the status of our human species that has ever occurred in history. You've now seen that either of the first two solutions to the paradox would have profound philosophical implications, to our place in the universe in the first case and to the prospects for our survival in the second. But the first solution seems extremely unlikely, and the second, while plausible, is one that we are rooting against. And that leaves us with the third solution: There *is* a galactic civilization.

Think about it. The favored solution to the "Where is everybody?" question is that they are here, all around us, but we are not yet capable of discovering their existence. If this is really true—and as incredible as it sounds, my own guess is that it is—it means that we are newcomers on the scene of a galactic civilization that has existed for millions or billions of years before us. Perhaps they have been deliberately hiding from us in some variation on the theme of Star Trek's "prime directive," in which the galactic federation has vowed not to interfere with emerging civilizations. Or perhaps they simply have no more reason to bother with us on our planet than we have to scoop up every handful of sand in a desert, and their technologies are so far ahead of us that we are not yet capable of detecting their presence. Whatever the reason why we have not yet found them, if they are real, it can mean only one thing: If we successfully navigate the adolescence of our civilization, there's a universe full of grownups awaiting our arrival.

BEYOND UFOS: THE TURNING POINT

Way back in the preface to this book, I told you about my tiny alien friends, the ones I once saw in beams of sunlight, but whom I have not seen for decades. Today, I suspect they were never really there on the motes of dust in the first place, and were only figments of my hopeful imagination. But, as you now see, I've traded my belief in aliens in the sunlight for something far more incredible. In leaving my mythical beings behind, I've found scientific reasons to think that there's a good chance, perhaps a very good chance, that we stand today on the verge of making contact with a civilization that predates ours by millions or billions of years. And even if this turns out not to be the case, and that one of the other solutions to the Fermi paradox is correct, we ourselves are positioned to become the galactic colonists, if only we can solve our current problems and mature as a species.

In essence, we find ourselves in a remarkable position in human history. Thousands of generations have come and gone, but none before us has ever had the ability to step off of this world into the great beyond, and to learn the answer to the age-old question of whether we are alone in the universe. However, along with this ability to advance our civilization, we have also developed the power to destroy. And thus, by accident of history, we are the generation that has been placed at the turning point. If we cannot redress our current failings, then we will bear the burden of having been the generation that ruined everything that our ancestors worked so hard to create. But if we can just grow up, the possibilities that await us are infinite.

Imagine for a moment the grand view, a gaze across the centuries and millennia from this moment forward. Picture our descendants living among the stars, having created or joined a great galactic civilization. They will have the privilege of experiencing ideas, worlds, and discoveries far beyond our wildest imaginations. Perhaps, in their history lessons, they will learn of our generation—the generation that history placed at the turning point, the generation that managed to steer its way past the dangers of self-destruction, and the generation that first stepped onto the path to the stars.

TO LEARN MORE

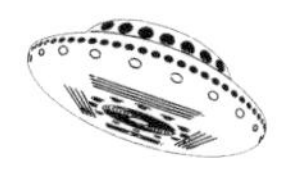

In a short book like this one, I've necessarily had to leave out many details that lie behind the topics discussed. I hope that some readers will be inspired to learn more, as well as to keep abreast of new developments in the fast-moving field of astrobiology. Toward that end, here are a few recommendations:

- You can find a vast amount of information on the Web, along with updates on new discoveries. A great place to start is the web site for the NASA Astrobiology Institute (nai.nasa.gov); after that, a search on "astrobiology" will take you to many other Web resources.
- If you would like more detail about the particular topics discussed in this book, you'll find them in my textbook *Life in the Universe* (currently in its second edition), co-authored by Seth Shostak and published by Addison Wesley.
- If you are a student or live in a college town, look for an introductory course in astrobiology or life in the universe; many institutions now offer such courses. (And if you teach at an institution that would like to start such a course, feel free to contact me and I'll help as much as I can.)
- Finally, while I won't make any promises, I plan to do my best to post updates and additional resources on the web site for this book: www.BeyondUFOs.com
- Please also visit my personal web site: www.JeffreyBennett.com
- Or Princeton University Press's web site for this book: press.princeton .edu/titles/8594.html

INDEX